HIGH FREQUENCY CONTINUOUS TIME FILTERS IN DIGITAL CMOS PROCESSES

HIGH FREQUENCY CONTINUOUS TIME FILTERS IN DIGITAL CMOS PROCESSES

Shanthi Pavan
Texas Instruments, Incorporated

Yannis Tsividis
Columbia University

KLUWER ACADEMIC PUBLISHERS
Boston / Dordrecht / London

Distributors for North, Central and South America:
Kluwer Academic Publishers
101 Philip Drive
Assinippi Park
Norwell, Massachusetts 02061 USA
Telephone (781) 871-6600
Fax (781) 871-6528
E-Mail <kluwer@wkap.com>

Distributors for all other countries:
Kluwer Academic Publishers Group
Distribution Centre
Post Office Box 322
3300 AH Dordrecht, THE NETHERLANDS
Telephone 31 78 6392 392
Fax 31 78 6546 474
E-Mail <orderdept@wkap.nl>

 Electronic Services <http://www.wkap.nl>

Library of Congress Cataloging-in-Publication Data

A C.I.P. Catalogue record for this book is available
from the Library of Congress.

Copyright © 2000 by Kluwer Academic Publishers

All rights reserved. No part of this publication may be reproduced, stored in
a retrieval system or transmitted in any form or by any means, mechanical,
photo-copying, recording, or otherwise, without the prior written permission
of the publisher, Kluwer Academic Publishers, 101 Philip Drive, Assinippi
Park, Norwell, Massachusetts 02061

Printed on acid-free paper.

Printed in the United States of America

Contents

List of Figures	ix
List of Tables	xix
Preface	xxi

1. INTRODUCTION ... 1
    1. MOTIVATION ... 1

2. MOS CAPACITOR MODELING ... 5
    1. INTRODUCTION ... 5
    2. DISTORTION GENERATION IN NONLINEAR CAPACITORS ... 6
    3. GATE CAPACITOR STRUCTURES IN CMOS TECHNOLOGY ... 8
    4. MODELING OF ACCUMULATION MOS CAPACITORS FOR ANALOG DESIGN ... 10
    5. SIMULATION RESULTS AND MODEL VALIDATION ... 13
    6. MODEL IMPLEMENTATION ... 15
    7. THE POLYSILICON GATE DEPLETION EFFECT ... 19
    8. MEASUREMENT RESULTS AND DISCUSSION ... 22
    9. HIGH FREQUENCY CHARACTERISTICS ... 23
    10. N-WELL TO SUBSTRATE CAPACITANCE ... 28
    11. SUMMARY ... 31

3. A REVIEW OF INTEGRATOR ARCHITECTURES ... 33
    1. INTRODUCTION ... 33
    2. NON-IDEALITIES IN INTEGRATORS ... 33
    3. MOSFET-C FILTERS ... 37
    4. Gm-C FILTERS ... 42
    5. Gm-OTA-C FILTERS ... 46
    6. A STUDY OF CMOS TRANSCONDUCTORS ... 48
    7. PROBLEMS OF PROGRAMMABLE FILTER DESIGN ... 56
    8. APPROACHES TO THE DESIGN OF WIDELY PROGRAMMABLE INTEGRATORS ... 60

9.	CONCLUSIONS	69

4. TIME SCALING IN ELECTRICAL NETWORKS — 71

1.	INTRODUCTION	71
2.	TIME SCALING: DEFINITION	72
3.	THE LINEAR CASE	72
4.	NOISE PROPERTIES OF SCALED NETWORKS	75
5.	EXTENSION OF SCALING TO THE NONLINEAR CASE	79
6.	DISTORTION IN WEAKLY NONLINEAR SCALED FILTERS	86
7.	IMPLEMENTATION OF SCALED INTEGRATORS IN CMOS VLSI	90
8.	CONCLUSIONS	93

5. FILTER DESIGN — 95

1.	INTRODUCTION	95
2.	FILTER DESIGN	95
3.	TRANSCONDUCTOR DESIGN	98
4.	COMMON-MODE FEEDBACK CIRCUIT	103
5.	FREQUENCY TUNING SYSTEM	109
6.	PARASITIC CAPACITANCES	116
7.	BIQUAD LEVEL LAYOUT	120
8.	SIMULATION RESULTS	123

6. FILTER TESTING AND MEASUREMENT RESULTS — 133

1.	INTRODUCTION	133
2.	FREQUENCY RESPONSE	136
3.	FILTER OUTPUT NOISE	138
4.	DISTORTION	143
5.	TEMPERATURE MEASUREMENTS	143
6.	SUMMARY	145

7. FURTHER APPLICATIONS OF SCALING — 149

1.	INTRODUCTION	149
2.	NAUTA'S CMOS VHF FILTER TECHNIQUE	149
3.	IMPROVED FILTER TECHNIQUE ...	152
4.	A MODIFIED Gm-OTA-C TECHNIQUE	157
5.	SUMMARY AND CONCLUSIONS	160

8. TUNING IN CONTINUOUS-TIME FILTERS — 165

1.	INTRODUCTION	165
2.	THE VOLTAGE CONTROLLED FILTER TECHNIQUE	167
3.	THE VOLTAGE CONTROLLED OSCILLATOR TECHNIQUE	170
4.	AN ANALYTICAL SOLUTION TO A CLASS OF OSCILLATORS	172
5.	APPLICATIONS OF THE FILTER COMPARATOR OSCILLATOR TO FILTER TUNING	182
6.	CONCLUSION	188

Appendices

A – AN EXPLICIT EXPRESSION FOR ψ_s IN A MOS ACCUMULATION CAPACITOR ... 199

B – CALCULATION OF THE BIAS VOLTAGE AT WHICH $\frac{dC_{GB}}{dV_{GB}} = 0$ ... 201

C – SUMMARY OF THE PROPERTIES OF A DISTRIBUTED RC LINE ... 205

D – SMALL SIGNAL MOS TRANSISTOR MODELS ... 207
    1.   OPERATION IN STRONG INVERSION AND SATURATION ... 207
    2.   MODEL WHEN THE DEVICE IS OFF ... 209
    3.   EXTRINSIC PARASITICS ... 209

E – CALCULATION OF THE STEADY STATE WAVEFORM OF THE FILTER COMPARATOR OSCILLATOR ... 211

Index ... 213

List of Figures

1.1	A hard disc platter.	2
1.2	The simplified block diagram of a read-channel chip.	2
2.1	Gm-C integrator with a non–linear capacitor.	6
2.2	Capacitor A generates a much smaller third harmonic than Capacitor B.	8
2.3	Gate capacitor structures in an n-well CMOS technology.	9
2.4	Cross section of an MOS capacitor.	11
2.5	Comparison of numerical simulation and proposed models for different values of substrate doping concentration (– numerical, ⊕ proposed). $C'_{ox} = 3.84$ fF/μm^2, $V_{FB} = 0$, T=300 K.	14
2.6	Variation of capacitance with temperature(– numerical, ⊕ proposed). $C'_{ox} = 3.84$ fF/μm^2, $V_{FB} = 0$, $N_D = 10^{17}$ cm^{-3}.	14
2.7	Circuit to compare both piecewise linear and explicit models.	16
2.8	Second harmonic relative to fundamental(– PWL, ⊕ proposed).	16
2.9	Third harmonic relative to fundamental(– PWL, ⊕ proposed).	17
2.10	Harmonics relative to fundamental for different values of bias voltage(– PWL, ⊕ proposed).	17

2.11	Qualitative C-V curves of a MOS capacitor with $n+$ and $p+$ polysilicon gates. Since an n-type substrate is assumed, accumulation is to the right of the minimum on each curve.	18
2.12	MOS capacitor cross section.	19
2.13	Comparison of complete model with measurements. $t_{ox} = 85$ Å (–model, $\oplus$ data).	23
2.14	Comparison of complete model with measurements. $t_{ox} = 45$ Å (–model, $\oplus$ data).	24
2.15	Structure of the MOS capacitor and derivation of a two dimensional $\overline{URY}$ equivalent circuit.	25
2.16	Distributed Model for the MOS capacitor.	26
2.17	Definitions of R_{ch} and R_g.	27
2.18	Distributed network to calculate $Y_c(s)$.	27
2.19	Part of a grid to simulate 2-D effects in SPICE.	29
2.20	Re(Y) computed using various models.	29
2.21	Im(Y) computed using various models.	30
2.22	Complete model for the poly-n-well MOS accumulation capacitor.	30
2.23	HD_2 as a function of bias voltage and signal level.	32
2.24	HD_3 as a function of bias voltage and signal level.	32
3.1	Magnitude and phase responses of an integrator	34
3.2	Pole forming section of a biquad	35
3.3	Pole forming section of a biquad with non-ideal integrators	35
3.4	A MOSFET-C integrator	37
3.5	Conceptual schematics of MOSFET-C integrators with accumulation capacitors.	39
3.6	A MOSFET-C integrator using accumulation capacitors.	40
3.7	An alternate MOSFET-C integrator using accumulation capacitors.	41
3.8	Two versions of the Gm-C integrator.	43
3.9	The Gm-C integrator with the common-mode setting loop explicitly shown.	44
3.10	A conceptual schematic of a Gm-OTA-C integrator.	46

3.11	Schematic of a Gm-OTA-C integrator with accumulation capacitors.	47
3.12	Alternative schematic of a Gm-OTA-C integrator with accumulation capacitors.	48
3.13	Another Gm-OTA-C integrator with accumulation capacitors.	48
3.14	A degenerated differential pair.	49
3.15	A degenerated differential pair with parasitic effects and excess noise sources.	50
3.16	The transconductor of Krummenacher and Joehl.	52
3.17	Transconductors based on opamps.	52
3.18	Conceptual schematic.	53
3.19	Realizations in CMOS and BiCMOS.	53
3.20	The differential pair.	54
3.21	Small signal single ended equivalent circuit of a differential pair.	55
3.22	Aberrations in filter response at high bandwidth settings due to variations in phase error.	59
3.23	Response of a programmable filter, where the integrators have a constant phase error, independent of their unity gain frequencies.	59
3.24	(a) A Gm-C integrator designed to achieve the maximum cutoff frequency in a programmable range; (b) Lowering the cutoff frequency of (a) by increasing the capacitance ("constant G_m" approach); (c) Lowering the cutoff frequency of (a) by decreasing the transconductance ("constant C" approach.)	61
3.25	Variation of mean squared noise, total capacitance and power dissipation versus cutoff frequency for programmable integrators using the "constant G_m" approach (thin line) and the "constant C" approach (thick line). The circle indicates a design optimized for the maximum f_c setting.	63
3.26	A conventional Gm-C integrator.	64
3.27	η and $\Delta\phi$ as a function of ω_o.	65

3.28	A programmable transconductor using multiple switchable unit transconductors.	66
3.29	(a) A second order Gm-C filter; (b) The filter in (a), made programmable by varying transconductances; the parasitic input and output capacitances also vary, as indicated by arrows; (c) The filter in (a), assuming all parasitics remain fixed while transconductances are varied.	68
4.1	Definition of time scaling.	72
4.2	An example network.	74
4.3	Illustration of constant-conductance and constant-capacitance scaling for the network of Figure 4.2.	75
4.4	Illustration of constant-conductance and constant-capacitance scaling for a general (trans)conductance-capacitance network.	75
4.5	Noise model for the conductances and transconductors.	76
4.6	Noise properties of constant-conductance scaled networks.	76
4.7	Noise properties of constant-capacitance scaled networks.	78
4.8	Model for nonlinear (trans)conductors and capacitors.	80
4.9	Constant-conductance scaling in nonlinear networks.	81
4.10	Constant-capacitance scaling in nonlinear networks.	84
4.11	A network example to illustrate scaling - (a) Original network, (b) network constant-C scaled by $\alpha = 2$.	85
4.12	Top: Output waveform $v_{oa}(t)$, Middle: Output waveform $v_{ob}(t)$, Bottom: $v_{oa}(t)$ and $v_{ob}(t/2)$ overlaid on the same plot.	86
4.13	Distortion in weakly nonlinear filters.	87
4.14	Distortion simulations of a scaled Butterworth filter.	88
4.15	Distortion simulations of a scaled Butterworth filter, plotted with the x-axis normalized to bandwidth.	89
4.16	A constant-conductance scaled integrator.	91
4.17	A digitally tuned variable capacitor.	91
4.18	A unit transconductance element.	92

List of Figures

4.19	The small-signal equivalent circuit of the transconductor of Figure 4.18 under differential excitation, for the cases (a) $b=1$ and (b) $b=0$ (where b denotes the state of the corresponding switches in Figure 4.18).	93
4.20	Parallel connection of two unit transconductors.	94
5.1	Block diagram of the test chip.	96
5.2	Biquad 1 used in the realization of the fourth order Butterworth filter prototype.	97
5.3	Biquad 2 used in the realization of the fourth order Butterworth filter prototype.	97
5.4	Unit transconductor cell.	97
5.5	Complete transconductor.	99
5.6	Offsets in a unit transconductance cell.	102
5.7	Block diagram of the common-mode feedback circuit.	103
5.8	A differential pair as a common-mode detector.	104
5.9	A degenerated differential pair as a common-mode detector.	105
5.10	Preliminary design of the CMFB loop.	106
5.11	Final CMFB loop.	107
5.12	Schematic of the bias generation circuit for the CMFB loops.	108
5.13	Conceptual schematic of a conventional resistor-servo loop.	110
5.14	Basic fixed transconductance bias circuit.	110
5.15	Deriving an improved fixed transconductance bias circuit.	112
5.16	Scheme with reduced sensitivity to the output resistance of V_x in Figure 5.15.	113
5.17	Circuit implementation of the scheme of Figure 5.16.	114
5.18	Complete "fixed transconductance bias" circuit, shown along with a transconductor used in the filter.	115
5.19	Percent g_m variation of the differential pair with temperature.	115
5.20	Equivalent representation of mobility reduction.	116
5.21	I_D as a function of V_{GS}.	117
5.22	g_m as a function of V_{GS}.	117

5.23	Cross section of an interconnect line.	118
5.24	Comparing the capacitance of a thin interconnect line for two different heights over a ground plane.	119
5.25	An inherently balanced interconnect technique.	121
5.26	Schematic and layout of the transconductor used in the filter.	121
5.27	Two transconductor layout representations, and their equivalent circuits in the presence of an oxide gradient.	122
5.28	Schematic of a biquadratic section.	123
5.29	Layout of a biquadratic section tolerant to small oxide gradients.	124
5.30	Frequency response as a function of programming word – linear scale.	125
5.31	Frequency response as a function of programming word – logarithmic scale.	126
5.32	Bandpass output of Biquad 1 as a function of programming word.	127
5.33	Filter response for temperature values of 25, 65 and 125°C.	128
5.34	Balance in the filter layout.	129
5.35	Output noise power spectral density plots as the frequency control word is stepped from 0000 thru 1111.	130
5.36	Integrated output noise as the frequency control word is stepped from 0000 thru 1111.	131
5.37	Results of Monte Carlo simulations of the filter with random offsets. The largest THD is 0.3%.	132
6.1	Simplified schematic of test setup.	133
6.2	Twoport representation of the test setup.	135
6.3	Simulation of the measurement technique when the filter bandwidth is set at the low end - the thin line is the response of the buffer path, the thick line is the actual filter response obtained by using (6.2). Since the buffer gain is almost flat in the filter passband, the response of the filter path coincides with the actual filter response and is not discernible on the plot.	136

List of Figures xv

6.4	Simulation of the measurement technique when the filter bandwidth is set at the high end - the thin lines are the responses of the buffer and filter paths, the thick line is the actual filter response obtained by using (6.2).	137
6.5	Simplified schematic of test buffer and output circuits.	138
6.6	Passband detail.	139
6.7	Frequency response.	139
6.8	Measured response compared with an ideal 4th order Butterworth response.	140
6.9	Filter response at $V_{dd} = 3\,V, 3.3\,V$ and $3.6\,V$.	141
6.10	Response of 20 filters measured under identical conditions. Mean bandwidth $= 59.97\,MHz$, Standard Deviation $0.5\,MHz$.	142
6.11	Output noise spectral density of the filter for different bandwidth settings.	142
6.12	THD as a function of input signal level ($f_{in} = f_{-3dB}/3$).	144
6.13	Frequency response at $T = 0°C, 45°C$ and $75°C$.	144
6.14	Percent deviation of filter bandwidth from the nominal value.	145
6.15	Chip Microphotograph.	146
7.1	Simplified schematic of Nauta's transconductance element	150
7.2	Complete schematic Nauta's transconductance element	151
7.3	Generation of the input common-mode level	151
7.4	Basic transconductor principle.	152
7.5	Unit transconductance element.	153
7.6	Equivalent circuits for the unit transconductance element for differential excitation when (a) $b = 0$ and (b) $b = 1$.	154
7.7	Complete unit element.	155
7.8	Generation of the input common-mode level.	155
7.9	Gate of a saturated device driven by a sinusoidal current.	156
7.10	Definition of currents and charges in the presence of varying terminal voltages. Lowercase symbols with capital subscripts denote total time-varying quantities.	156

7.11	Magnitude of gate, bulk and inversion charges as a function of gate voltage ($v_D = 3\,\text{V}, v_S = v_B = 0\,\text{V}, W = 4.5\,\mu\text{m}, L = 1\,\mu\text{m}, t_{ox} = 48\,\text{Å}$.)	157
7.12	Gate capacitance as a function of v_{GS}.	158
7.13	Distortion in the gate capacitance voltage waveform.	158
7.14	A Gm-OTA-C integrator.	159
7.15	Scalable transconductor and OTA unit cells.	161
7.16	A digitally programmable transconductor and OTA.	162
7.17	A constant-capacitance scaled Gm-OTA-C integrator.	163
8.1	Concept of direct tuning.	166
8.2	Concept of indirect tuning.	167
8.3	A Gm-C biquad realizing lowpass and bandpass transfer functions.	168
8.4	A conventional Vector Lock Loop(VLL).	169
8.5	(a) ϕ and (b) M surfaces for a conventional VLL (frequency normalized to reference.)	170
8.6	Block diagram of a VCO tuning loop.	171
8.7	Block diagram of an oscillator.	172
8.8	The filter comparator oscillator.	173
8.9	Block diagram of the oscillator.	174
8.10	Oscillation buildup mechanism.	176
8.11	Timing detail.	178
8.12	Circuit schematic.	180
8.13	Filter and comparator outputs.	181
8.14	Measured($\oplus$) and predicted(—) amplitude of oscillation vs. Q.	182
8.15	Measured($\oplus$) and predicted(—) frequency of oscillation vs. Q.	183
8.16	Measured($\oplus$) and predicted(—) THD vs. Q.	183
8.17	Comparison of conventional and proposed technique.	184
8.18	Proposed VLL.	186
8.19	(a) Frequency and (b) Amplitude surfaces for proposed VLL (frequency normalized to desired pole frequency.)	189
8.20	Functionality testing of the proposed VLL.	190
B.1	Simulation of the effect of finite polysilicon doping on C–V characteristics.	202

C.1	A uniformly distributed RY line.	205
C.2	Admittance of a $\overline{URY}$ line contacted at both ends.	206
D.1	The intrinsic part of an MOS transistor.	208
D.2	A simple quasistatic model for the MOS transistor.	208
D.3	MOS transistor model in the off condition.	209
D.4	Extrinsic parasitic capacitors near the drain.	210
D.5	Extrinsic transistor capacitances added to an intrinsic model.	210

List of Tables

4.1	Basic properties of scaled networks, $\mathcal{N} \xrightarrow{t|\alpha t} \hat{\mathcal{N}}$.	73
4.2	Noise properties of constant-conductance scaled networks, $\mathcal{N} \xrightarrow{t|\alpha t} \hat{\mathcal{N}}$.	77
4.3	Noise properties of constant-capacitance scaled networks, $\mathcal{N} \xrightarrow{t|\alpha t} \hat{\mathcal{N}}$.	79
5.1	Center frequencies and quality factors of the individual biquads in a fourth order Butterworth filter with a bandwidth of ω_o.	96
6.1	Summary of measured characteristics (25°C, unless noted otherwise)	147

Preface

There is an ever increasing trend towards putting entire systems on a single chip. This means that analog circuits will have to coexist on the same substrate along with massive digital systems. Since technologies are optimized with these digital systems in mind, designers will have to make do with standard CMOS processes in the years to come. We address analog filter design from this perspective. Filters form important blocks in applications ranging from computer disc-drive chips to radio transceivers.

In this book, we develop the theory and techniques necessary for the implementation of high frequency (hundreds of megahertz) programmable continuous time filters in standard CMOS processes. Since high density poly-poly capacitors are not available in these technologies, alternative capacitor structures have to be found. Metal-metal capacitors have low specific capacitance. An alternative is to use the (inherently nonlinear) capacitance formed by MOSFET gates. In Chapter 2, we focus on the use of MOS capacitors as integrating elements. A physics-based model which predicts distortion accurately is presented for a two-terminal MOS structure in accumulation. Distortion in these capacitors as a function of signal swing and bias voltage is computed.

Chapter 3 reviews continuous-time filter architectures in the light of bias-dependent integrating capacitors. We also discuss the merits and demerits of various CMOS transconductance elements. The problems encountered in designing high frequency *programmable* filters are discussed in detail.

Chapter 4 considers time-scaling in electrical networks and introduces a technique called constant-capacitance scaling. We show that wide-range programmable filters implemented using this technique are optimal with respect to noise and dynamic range.

Chapter 5 documents the detailed design and layout of a 60 − 350 MHz Butterworth filter implemented in a 0.25 μm digital CMOS process. A simple circuit arrangement is proposed to keep the filter bandwidth constant with respect to temperature and process variations. Simulation results for the filter test chip are shown.

For filters in the frequency range of interest to us, special care has to be taken to measure the filter response accurately. In Chapter 6, we present measurement techniques and the implementation results of the prototype filter chip. This experimental data demonstrates the effectiveness of the techniques proposed in Chapter 4.

In Chapter 7, we discuss the application of scaling techniques to other filter architectures.

For very high-quality filters, the simple tuning technique proposed in Chapter 5 may not guarantee sufficient precision. In Chapter 8, a more accurate and complex method based on the behavior of a filter–comparator oscillator is proposed. This scheme converts a filter into an oscillator. The amplitude and frequency of oscillation are measures of the center frequency and quality factor of a biquadratic filter section. Analytical relations for the behavior of the filter–comparator system are developed and verified through a breadboard prototype.

This book is based on a doctoral thesis (submitted by one of the authors [S.P.] to Columbia University). It describes the results of a research project carried out to determine if it is at all possible to realize robust, production quality VHF filters in standard deep submicron CMOS technologies. The filter design discussed in this book was implemented at Texas Instruments (New Jersey).

ACKNOWLEDGMENTS: We are grateful to the staff of Texas Instruments at Warren for their support - T.R. Viswanathan and K. Nagaraj have been sources of great encouragement all along. We thank the students of the Columbia Integrated Systems Laboratory - especially M. Tarsia, G. Palaskas, N. Krishnapura and A. Dec for useful discussions. Finally, one of the authors [S.P.] would like to thank Professors Anthony Reddy and John Khoury - their superb classes on filter theory and circuit design have been profoundly influential over the course of creating this book.

SHANTHI PAVAN

YANNIS TSIVIDIS

To my parents ...

S.P.

Chapter 1

INTRODUCTION

1. MOTIVATION

High-speed communication and storage systems need continuous-time filters with bandwidths tunable over a wide range, while keeping the relative shape of the frequency response identical irrespective of the set bandwidth, and maintaining an adequate dynamic range. Typical system examples of such communication systems include magnetic storage (disc and tape drives), optical storage (CD-ROM drives) and high speed local area networks (LAN). A specific system example has been chosen below in order to emphasize the requirements and challenges of CMOS filter design at very high frequencies.

The disc-drive read-channel industry has been the driving force for realizing high-frequency broadband filters programmable over a wide range. A simplistic view of the read-channel system is given here [1]. A diagram of a hard disc platter is shown in Figure 1.1. The platter (coated with a magnetic material) rotates with a constant angular velocity which depends on details of mechanical construction and tolerances. Reading and writing onto the disc is accomplished by means of a read/write head located on the tip of the movable arm. (The motion of the arm itself is controlled by a complex mixed-signal chip, with a separate digital signal processor - however that is not the focus here.) The signal from the read head is the input for the read-channel chip, whose simplified block diagram is shown in Figure 1.2. The channel consists of a variable gain amplifier(VGA), to tune

2 HIGH FREQUENCY CONTINUOUS TIME FILTERS

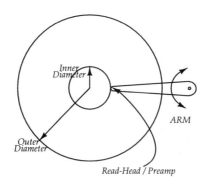

Figure 1.1. A hard disc platter.

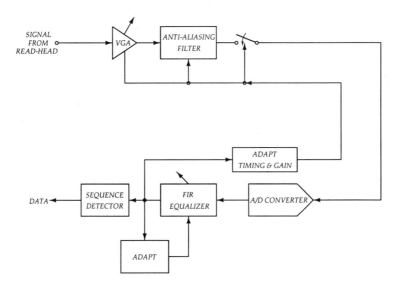

Figure 1.2. The simplified block diagram of a read-channel chip.

out any amplitude variations in the signal. A continuous-time filter (CTF) is the next block in the signal chain. Its function is two fold. First, it acts as an anti-aliasing filter, to prepare for signal digitization downstream. It also serves as a first pass equalizer for the channel. The signal is sampled and quantized in an analog-to-digital converter and fed into a custom DSP, which uses sophisticated signal processing techniques to detect the digital sequence written on to the platter. In order to optimally use the capacity of the disc, data is recorded with a uniform density on the platter. Since the *angular velocity* is a constant, it follows that the data-rate at the outer diameter is higher than that at the inner diameter. In other words, the bandwidth of the signal varies as the arm progresses from the inner to the outer diameter of the platter. For best immunity to noise, the bandwidth of the CTF has to be varied with the position of the arm. Typically the bandwidth must be programmable by at least a factor of three. Since process tolerances and temperature variations will also cause changes in bandwidth, the filter must have a larger raw tunable range. Moreover, there are tight specifications on the response of the filter as it is tuned from the lowest to highest bandwidth. Thankfully for the filter designer, the desired dynamic range is modest (35 − 40 dB). The most challenging part of the filter design is to *maintain frequency response accuracy and dynamic range across such a large tuning range.*

As can be inferred from the Figure 1.2, the predominant part of the channel is digital in nature. Hence, CMOS technology is the obvious choice. The trend in VLSI is towards smaller channel lengths and lower supply voltages. This increases the operating speed of the digital circuits, while lowering their power dissipation and shrinking their area. Cost and performance are extremely critical in a high-volume market such as this, and this has forced the industry to use a low-cost digital CMOS process. The factors cited above present serious problems for the analog part of the chip for the following reasons:

- Lower supply voltages mean lower signal swings and reduced dynamic range.
- Substrate coupling increases because the digital portion switches at much higher speeds.
- The output conductances of the devices are large in short-channel technologies - reducing the open loop gains achievable in amplifiers.
- High quality poly-poly capacitors, crucial to the performance of many analog circuits, are not available.

This book addresses the theoretical and practical problems encountered in the design of robust, programmable continuous-time filters with very high bandwidths (hundreds of megahertz) implemented in low cost digital CMOS technologies.

Fundamental work on the design of integrated continuous-time filters dates back to the seventies [2] [3]. For a review of the work done prior to 1990, the reader is referred to [4]. Up to that time, the emphasis was mostly on design techniques for fixed frequency filters. Tunability was considered mostly from the viewpoint of being able to maintain a stable response over process and temperature. The demand for *programmable* filters increased in a big way since the emergence of the disk-drive market, and several high frequency programmable filters were described in the nineties; see, for example, references [5]-[12].

The filter examples and architectures cited above are only representative of what has been published. A careful look at the techniques presented in the literature reveals that the state of the art in CMOS filters is lacking in performance when compared to realizations in BiCMOS. The methods proposed in the literature cannot be "pushed" to frequencies in the hundreds of megahertz range; for example, it would be very difficult to implement a 1.8 V, 50-500 MHz, 60 dB dynamic range standard CMOS filter using conventional techniques. A fresh look at the problem from the viewpoint of programmability is necessary.

Chapter 2

MOS CAPACITORS : MODELING & POTENTIAL AS INTEGRATING ELEMENTS IN CONTINUOUS TIME FILTERS

1. INTRODUCTION

In this chapter, we investigate the use of MOS capacitors as integrating elements in continuous time filters. The motivation for this study is to be able to design filters in low cost digital CMOS processes where high density poly-poly capacitors are not available. Although we could use metal-metal capacitors, the area required to realize a given capacitance is prohibitively large. An alternative is to use the capacitance formed by MOSFET gates. Gate oxide thickness is typically a tightly controlled variable. Moreover, MOS structures are naturally available in all CMOS technologies. The advantages of using the gate oxide capacitance are precision in absolute value and large specific capacitance. However, these capacitors are inherently nonlinear. We investigate the limits on the voltage swing across the capacitors for a given level of distortion.

First we evaluate distortion generated in a Gm-C integrator when the transconductor is ideal but the capacitor is weakly nonlinear. Next we examine the various operating regions possible on the C-V characteristic of a two terminal MOS structure. We will show that operating the capacitors in accumulation is preferable to operation in inversion. Then, we propose an explicit analytical model for the capacitance of a two terminal MOS structure in accumulation. This is useful for implementation in a general purpose circuit simulator like SPICE. Finally parasitics, high frequency effects and capacitor layout are considered.

6 HIGH FREQUENCY CONTINUOUS TIME FILTERS

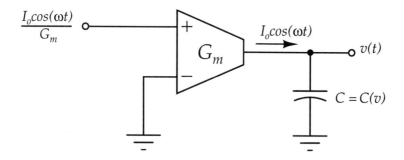

Figure 2.1. Gm-C integrator with a non–linear capacitor.

2. ANALYSIS OF DISTORTION GENERATION IN NONLINEAR CAPACITORS

For a nonlinear capacitor, the relation between charge and incremental capacitance can be written as

$$C(v) = \frac{dQ(v)}{dv} \qquad (2.1)$$

If the capacitor has a voltage $v(t)$ across it,

$$i = \frac{dQ(v)}{dv}\frac{dv}{dt} = C(v)\frac{dv}{dt} \qquad (2.2)$$

Consider now the Gm-C integrator shown in Figure 2.1. Let the transconductor be ideal and the initial voltage on the capacitor be V_o. Since the capacitor is weakly nonlinear, we assume that it is sufficiently accurately modeled (around the operating point ($v = V_o$)) by

$$C(v) = C_0 + C_1(v - V_o) + C_2(v - V_o)^2 \qquad (2.3)$$

The current flowing into the capacitor is sinusoidal with amplitude I_o. Hence, we have

$$I_o \cos(\omega t) = (C_0 + C_1(v - V_o) + C_2(v - V_o)^2)\frac{dv}{dt} \qquad (2.4)$$

Integrating with respect to time on both sides of the above, and using $v(0) = V_o$, we obtain

$$I_o \frac{\sin(\omega t)}{\omega} = C_0(v - V_o) + C_1\frac{(v - V_o)^2}{2} + C_2\frac{(v - V_o)^3}{3} \qquad (2.5)$$

Solving the differential equation (2.4) in general can be very complicated. Since the capacitor is weakly non-linear, we use an approximate method to determine distortion components in the capacitor voltage. $C_1(v - V_o)^2$ and $C_2(v - V_o)^3$ are assumed to be small compared to $C_o(v - V_o)$. Then, we can invert (2.5) as

$$v - V_o \approx I_o \frac{\sin(\omega t)}{\omega C_0} - \frac{C_1}{2C_0}\left[I_o \frac{\sin(\omega t)}{\omega C_0}\right]^2 - \frac{C_2}{3C_0}\left[I_o \frac{\sin(\omega t)}{\omega C_0}\right]^3 \quad (2.6)$$

It can be easily verified that $v(t)$ can be written as

$$v(t) = A + B\sin(\omega t) + C\cos(2\omega t) + D\sin(3\omega t) \quad (2.7)$$

where

$$A \approx V_o$$
$$B \approx \frac{I_o}{\omega C_0}$$
$$C \approx \frac{C_1}{4C_0} B^2$$
$$D \approx \frac{C_2}{12C_0} B^3$$

Hence,

$$HD_2 \approx \frac{C_1}{4C_0} B \quad (2.8)$$

$$HD_3 \approx \frac{C_2}{12C_0} B^2 \quad (2.9)$$

Note that although the above analysis is sufficient to estimate distortion, it *does not* predict the small change in the fundamental due to higher order non-linearities.

If fully differential circuits are used, even order distortion is eliminated. The limiting factor, then, is third harmonic distortion. (2.3), (2.8) and (2.9) can be physically interpreted as follows: Second harmonic distortion is determined by the *slope* of the C–V curve around the operating point; and third harmonic distortion is determined by the *curvature* of the C–V curve around the operating point. These observations can be used to conclude that, for example, in Figure 2.2, we would choose capacitor A to design a fully differential Gm-C integrator because it generates smaller third harmonic distortion

8 HIGH FREQUENCY CONTINUOUS TIME FILTERS

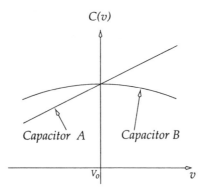

Figure 2.2. Capacitor A generates a much smaller third harmonic than Capacitor B.

compared to capacitor B, *even though* B's C–V curve *looks much flatter*. This discussion brings to light yet another point. Any models used for circuit design should model derivatives accurately, if one is interested in predicting distortion through simulation.

3. GATE CAPACITOR STRUCTURES IN CMOS TECHNOLOGY

Figure 2.3 shows four MOS capacitor structures in an *n*-well CMOS technology. The maximum achievable specific capacitance is that corresponding to the thickness of the gate oxide. The top plate of the gate-oxide capacitor is formed by polysilicon. The bottom plate must be contacted through the source/drain regions. To simplify the discussion that follows, assume that the magnitude of the flatband voltage is small. First, let us consider an *n*-type substrate. For the structure of Figure 2.3(a), if a large positive voltage is applied to the top plate, positive charges on the polysilicon attract negative charges in the bulk to the oxide-semiconductor interface, leading to accumulation of electrons at the surface. The bottom plate contact is formed by the *n*+ diffusion. The structure of Figure 2.3(b) can be used if the bottom plate is formed by holes. This corresponds to a negative gate-body voltage and an inverted semiconductor surface. We conclude that the structure of Figure 2.3(a) is suitable for accumulation mode of operation, while that in (b) is to be operated in inversion. Similar arguments can be made for the capacitors in (c) and (d). If we are only interested in positive values for V_{gb}, we

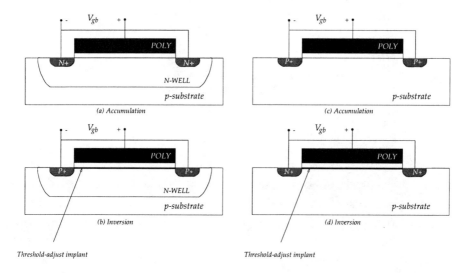

Figure 2.3. Gate capacitor structures in an n-well CMOS technology.

are left with two choices among the four structures discussed. The structure of (a) must be operated in accumulation; alternatively, the structure of (d) must be operated in inversion.

A comparison of distortion performance between accumulation and inversion capacitors is derived in [14]. Here, we summarize the results of that work. For large bias voltages ($V_{gb} > 2V$),

$$C'_{inv} = C'_{ox}\left[1 - \frac{2\phi_t}{V_{gb} - V_T + 2\phi_t}\right] \quad (2.10)$$

$$C'_{acc} = C'_{ox}\left[1 - \frac{2\phi_t}{V_{gb} - V_{FB} + 2\phi_t}\right] \quad (2.11)$$

where the primes denote capacitances per unit area, V_T is the threshold voltage of the NMOS structure of Figure 2.3(d), V_{FB} is the flatband voltage of the structure of Figure 2.3(a), and ϕ_t is the thermal potential (kT/q). If the gate polysilicon is n-type, V_{FB} is almost zero, and hence, C'_{acc} is much closer to C'_{ox} than C'_{inv}. By expanding (2.11) in a Taylor series around the bias point V_{gb}, and using the results of Section 2, we get

$$HD_{2,inv} = \frac{\phi_t V_p}{2(V_{gb} - V_T)^2} \quad (2.12)$$

$$HD_{2,acc} = \frac{\phi_t V_p}{2(V_{gb} - V_{FB})^2} \tag{2.13}$$

$$HD_{3,inv} = \frac{\phi_t V_p^2}{6(V_{gb} - V_T)^3} \tag{2.14}$$

$$HD_{3,acc} = \frac{\phi_t V_p^2}{6(V_{gb} - V_{FB})^3} \tag{2.15}$$

where V_p denotes the amplitude of the voltage across the capacitor. From the equations above, and assuming $V_{gb} - V_T << V_{gb} - V_{FB}$ we see that the distortion generated by a capacitor operating in inversion is much higher than that generated by one operating in accumulation. This is the motivation for concentrating on accumulation capacitors. An interesting observation is that V_T could vary significantly among different technologies, while V_{FB} has a smaller variation. Thus, the properties of accumulation capacitors are more "universal".

The results in [14] are valid for a relatively large bias compared to today's voltages. Since the trend in VLSI is definitely towards lower supplies, careful modeling of MOS accumulation capacitors valid even at low voltages is necessary. As pointed out in Section 2, to determine distortion accurately by simulation, the *derivatives* of the C–V curve need to be captured accurately in the model. This is the subject of the rest of this chapter [15].

4. MODELING OF ACCUMULATION MOS CAPACITORS FOR ANALOG DESIGN

Consider the two–terminal poly-n-well MOS structure shown in Figure 2.4. For a given bias voltage V_{GB} applied across the device, we derive relations for charge, capacitance and surface potential from first principles. For a detailed discussion of MOS device behavior, the reader is referred to [16].

ψ_{ox} is the potential drop across the oxide. ψ_s is the surface potential. ϕ_{MS} is the potential corresponding to the work function difference of the bulk and the gate materials. By Kirchhoff's voltage law,

$$V_{GB} = \phi_{MS} + \psi_{ox} + \psi_s \tag{2.16}$$

In the analysis to follow, Q'_G and Q'_c denote the charge per unit area on the gate and in the semiconductor respectively. Q'_o is the effective

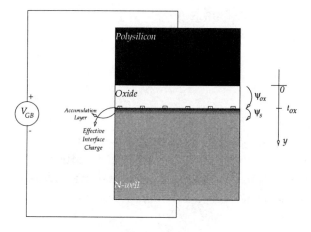

Figure 2.4. Cross section of an MOS capacitor.

interface charge per unit area of the oxide. ϵ_s, N_D and ϕ_t stand for the dielectric constant of silicon, donor density in the *n*-well and the thermal voltage respectively. C'_{ox} is the capacitance of the oxide per unit area and ϕ_F is the Fermi potential of the *n*-well.

Charge balance leads to

$$Q'_G + Q'_c + Q'_o = 0 \tag{2.17}$$

ψ_{ox} can be written as

$$\psi_{ox} = \frac{Q'_G}{C'_{ox}} = -\frac{Q'_c + Q'_o}{C'_{ox}} \tag{2.18}$$

Using the above for ψ_{ox} in (2.16), we obtain

$$V_{GB} = V_{FB} + \psi_s - \frac{Q'_c}{C'_{ox}} \tag{2.19}$$

where V_{FB} is the flatband voltage of the two terminal structure and is given by

$$V_{FB} = \phi_{MS} - \frac{Q'_o}{C'_{ox}} \tag{2.20}$$

Solving Poisson's Equation in the y direction, we obtain the total charge per unit area in the semiconductor as [16]

$$Q'_c = -\sqrt{2q\epsilon_s N_D}\sqrt{\phi_t \exp(\frac{\psi_s}{\phi_t}) - \psi_s - \phi_t + \left[\exp(\frac{2\phi_F}{\phi_t})\left(\phi_t \exp(\frac{-\psi_s}{\phi_t}) + \psi_s - \phi_t\right)\right]} \quad (2.21)$$

In accumulation, $\psi_s > 0$. Notice that ϕ_F is negative for an n-type substrate. For typical values of N_D and temperature, in accumulation, the quantity in square brackets in (2.21) will be orders of magnitude smaller than the other terms. Thus, it can be neglected, and (2.21) can be simplified as

$$Q'_c = -\sqrt{2q\epsilon_s N_D}\sqrt{\phi_t \exp(\frac{\psi_s}{\phi_t}) - \psi_s - \phi_t} \quad (2.22)$$

Notice that (2.22) is equivalent to assuming that there are no holes in the n-well. It can also be shown that the concentration of holes can be neglected while calculating the capacitance of the entire structure. From now on, therefore, we will use (2.22) rather than (2.21) in our analysis.

Using (2.22) with (2.19) we get

$$V_{GB} = V_{FB} + \psi_s + \gamma\sqrt{\phi_t \exp(\frac{\psi_s}{\phi_t}) - \psi_s - \phi_t} \quad (2.23)$$

where γ is given by

$$\gamma = \frac{\sqrt{2q\epsilon_s N_D}}{C'_{ox}} \quad (2.24)$$

For a given bias V_{GB}, we can solve (2.23) numerically for ψ_s and calculate Q'_c using (2.22). However, if an explicit expression for the surface potential can be obtained, calculations become much more straightforward. It can be found that, for typical values of γ and temperature, such an explicit expression is (see Appendix A),

$$\psi_s = 2\phi_t \left[\frac{V_{GB} - V_{FB} + k_1\phi_t}{V_{GB} - V_{FB} + k_2\phi_t}\right] \log\left(1 + \frac{V_{GB} - V_{FB}}{\gamma\sqrt{\phi_t}}\right) \quad (2.25)$$

where $k_1 = 3$ and $k_2 = 6$ provide a good fit to a numerical solution of (2.23), for a wide range of process parameters.

If the voltage across the two terminal structure is increased by ΔV_{GB}, by potential balance,

$$\Delta V_{GB} = \Delta \psi_{ox} + \Delta \psi_s \qquad (2.26)$$

Dividing by the charge per unit area $\Delta Q'_G$ which flows into the gate as a result of this voltage increase,

$$\frac{\Delta V_{GB}}{\Delta Q'_G} = \frac{\Delta \psi_{ox}}{\Delta Q'_G} + \frac{\Delta \psi_s}{\Delta Q'_G} = \frac{\Delta \psi_{ox}}{\Delta Q'_G} + \frac{\Delta \psi_s}{-\Delta Q'_c} \qquad (2.27)$$

or, allowing the finite differences to approach zero,

$$\frac{1}{C'_{gb}} = \frac{1}{C'_{ox}} + \frac{1}{C'_c} \qquad (2.28)$$

where $C'_c = -\frac{dQ'_c}{d\psi_s}$ is the capacitance of the accumulation layer per unit area. Differentiating (2.22),

$$C'_c = \gamma C'_{ox} \frac{\exp(\frac{\psi_s}{\phi_t}) - 1}{2\sqrt{\phi_t \exp(\frac{\psi_s}{\phi_t}) - \psi_s - \phi_t}} \qquad (2.29)$$

To summarize, for a given V_{GB}, the capacitance of the two terminal MOS structure can be found using the following three steps.

- Step 1: Solve (2.23) numerically for ψ_s. Alternatively, use the (approximate) explicit relation (2.25) to find ψ_s.
- Step 2: Use the value of ψ_s obtained in Step 1 in (2.29) to calculate C'_c.
- Step 3: Use C'_c calculated in Step 2 in (2.28) to obtain C'_{gb}.

5. SIMULATION RESULTS AND MODEL VALIDATION

In this section, we compare the explicit model using (2.25) proposed in the previous section with numerical simulations. Figure 2.5 compares the capacitance obtained by numerical methods with the explicit model proposed in this thesis. C'_{ox} was kept constant and the N_D values used were 10^{16}cm^{-3}, 10^{17}cm^{-3} and 10^{18}cm^{-3}. Note that this corresponds to the variation of γ by a factor of ten. Figure 2.6 compares the C–V characteristics as temperature is swept over a 150 degree range.

14 HIGH FREQUENCY CONTINUOUS TIME FILTERS

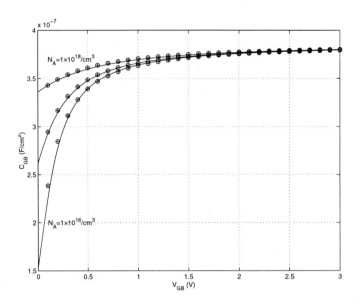

Figure 2.5. Comparison of numerical simulation and proposed models for different values of substrate doping concentration (– numerical, ⊕ proposed). $C'_{ox} = 3.84$ fF/μm^2, $V_{FB} = 0$, $T=300$ K.

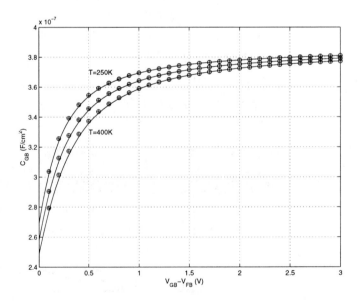

Figure 2.6. Variation of capacitance with temperature(– numerical, ⊕ proposed). $C'_{ox} = 3.84$ fF/μm^2, $V_{FB} = 0$, $N_D = 10^{17}$ cm^{-3}.

6. MODEL IMPLEMENTATION

In this section we examine issues that arise in implementing the model in a general purpose circuit simulator. In this work, the model was implemented in TISPICE, an in-house circuit simulation program at Texas Instruments. The model is implemented as a user-defined utility, coded as a 'C' routine. The inputs to the routine are the gate bias, flatband voltage, temperature, area and γ. The outputs are capacitance (C_{gb}) and charge (Q_G) on a capacitor plate for the given bias. C_{gb} and Q_G are obtained by multiplying C'_{gb} and Q'_G with the area of the capacitor respectively.

C'_{gb} is calculated using (2.28). However, one needs to be careful about the manner in which Q'_G is obtained. There are two possibilities for computing Q'_G. One could use (2.22) to obtain

$$Q'_G = -Q'_c - Q'_o = \sqrt{2q\epsilon_s N_D} \sqrt{\phi_t \exp(\frac{\psi_s}{\phi_t}) - \psi_s - \phi_t} - Q'_o \quad (2.30)$$

Alternatively, from (2.18) and (2.16)

$$Q'_G = C'_{ox}\psi_{ox} = C'_{ox}(V_{GB} - \psi_s - \phi_{MS}) \quad (2.31)$$

The two approaches would yield identical results *if ψ_s was known exactly*. However, our expression (2.25) for ψ_s is approximate. When the applied gate bias is large, even a small error in determining the surface potential would lead to a large error in computing charge from (2.30). On the other hand, (2.31) has no such problem. Hence, this expression should be used to calculate capacitor charge.

The performance of the implemented model was compared with a very finely spaced piecewise linear capacitor model. This model was generated from device simulation. The voltage step used in the piecewise linear model was 1 mV. Distortion is used as a metric for comparison, as this depends on *derivatives* of the C–V characteristic, and is hence a very sensitive indicator of modeling accuracy.

The SPICE test circuit that was used to compare distortion is shown in Figure 2.7. Both capacitors are driven from the same voltage source, and the distortion components of the currents are compared. Since third harmonic distortion is very small at high bias voltages and/or low signal levels, simulations were run at very tight tolerances, and using a very small time step compared to the period

16   HIGH FREQUENCY CONTINUOUS TIME FILTERS

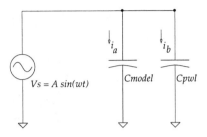

Figure 2.7. Circuit to compare both piecewise linear and explicit models.

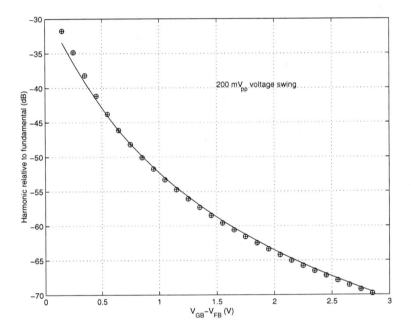

Figure 2.8. Second harmonic relative to fundamental(− PWL, ⊕ proposed).

of the voltage source. Putting both capacitors in the same circuit ensures that the time steps used in the calculation of either capacitor currents are identical. The dynamic range of the simulations is about 100 dB. Figure 2.8 and 2.9 compare the levels of second and third harmonic currents in the capacitors. Figure 2.10 shows variation in distortion components as a function of signal level for two different values of capacitor bias.

MOS Capacitor Modeling 17

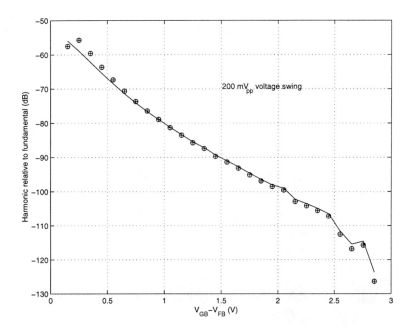

Figure 2.9. Third harmonic relative to fundamental(– PWL, ⊕ proposed).

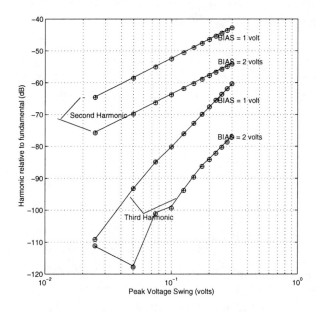

Figure 2.10. Harmonics relative to fundamental for different values of bias voltage(– PWL, ⊕ proposed).

18   HIGH FREQUENCY CONTINUOUS TIME FILTERS

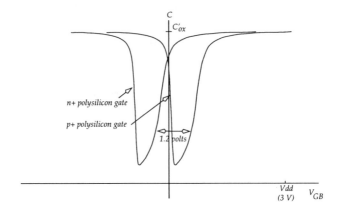

Figure 2.11. Qualitative C-V curves of a MOS capacitor with $n+$ and $p+$ polysilicon gates. Since an n-type substrate is assumed, accumulation is to the right of the minimum on each curve.

6.1 CIRCUIT CONSIDERATIONS AND CHOICE OF POLYSILICON TYPE

Until now, the results we derived for Q'_G, Q'_c and ψ_s were independent of the type of polysilicon used as the gate material. We now examine the two choices for the the type of polysilicon (n or p) from a circuit design viewpoint. As mentioned in the introduction to this paper, we are interested in using MOS accumulation capacitors in place of metal-metal capacitors. We would like to operate the structure in deep accumulation even for a very small V_{GB}. Hence, the flatband voltage (V_{FB}) should be as negative as possible. In modern CMOS technologies C'_{ox} is very large, and the term $\frac{Q'_o}{C'_{ox}}$ in (2.20) is typically about 100 mV. If the gate is degenerately doped (either $n+$ or $p+$), ϕ_{MS} is given by

$$\phi_{MS} = \begin{cases} -\phi_F - 0.56\text{V} , & n+ \text{ poly} \\ -\phi_F + 0.56\text{V} , & p+ \text{ poly} \end{cases} \quad (2.32)$$

The C-V curves for the cases when the gate polysilicon is doped $n+$ and $p+$ are shown in relation to the supply voltage in Figure 2.11. Note that since an n-type substrate is assumed, accumulation is to the right of the minimum on each curve. From this figure and the equation above, $n+$ polysilicon is the proper choice. We will henceforth assume that an $n+$ poly gate is used.

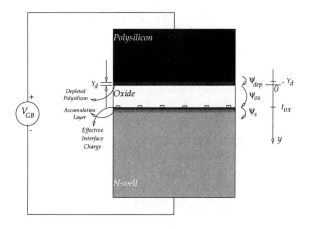

Figure 2.12. MOS capacitor cross section.

7. THE POLYSILICON GATE DEPLETION EFFECT

The analysis presented in the preceding sections assumes that the "top plate" of the MOS capacitor is infinitely highly doped. In practice, the finite doping density of the $n+$ polysilicon causes a depletion layer in the gate, whose thickness depends on the amount of negative charge in the n-well. In the literature[17][18], the effects of gate depletion are analyzed for the case when the MOS structure is in inversion. In such work, accumulation is not considered as this region is not important for transistor operation.

We redraw the cross section of the MOS capacitor in Figure 2.12, now with a depletion layer in the gate. Assume that the gate is uniformly doped n-type with a donor concentration of N_{POLY}. The figure is not to scale – the thickness of the gate depletion layer (denoted by Y_d in the diagram) and the accumulation layer have been grossly exaggerated for clarity. Let ψ_{dep} denote the potential drop across the gate depletion layer. Let a voltage V_{GB} be applied to the structure as shown. The potential balance equation can be written as

$$V_{GB} = \phi_{MS} + \psi_{dep} + \psi_{ox} + \psi_s \quad (2.33)$$

Charge balance, as given by (2.17), still holds. Using (2.18) in (2.33), we get

$$V_{GB} = \phi_{MS} + \psi_{dep} - \frac{Q'_c + Q'_o}{C'_{ox}} + \psi_s \quad (2.34)$$

Using (2.20), the above equation can be rewritten as

$$V_{GB} = V_{FB} + \psi_{dep} + \psi_s - \frac{Q'_c}{C'_{ox}} \quad (2.35)$$

Q'_c is only a function of ψ_s and is given by (2.22). Using (2.22) with (2.35)

$$V_{GB} = V_{FB} + \psi_s + \psi_{dep} + \gamma\sqrt{\phi_t \exp(\frac{\psi_s}{\phi_t}) - \psi_s - \phi_t} \quad (2.36)$$

From basic electrostatics,

$$Q'_G = \sqrt{2q\epsilon_s N_{POLY}}\sqrt{\psi_{dep}} \quad (2.37)$$

From (2.18) and (2.33)

$$Q'_G = C'_{ox}(V_{GB} - \phi_{MS} - \psi_s - \psi_{dep}) \quad (2.38)$$

Hence,

$$\sqrt{2q\epsilon_s N_{POLY}}\sqrt{\psi_{dep}} = C'_{ox}(V_{GB} - \phi_{MS} - \psi_s - \psi_{dep}) \quad (2.39)$$

Analogous with (2.24), we define

$$\gamma_p = \frac{\sqrt{2q\epsilon_s N_{POLY}}}{C'_{ox}} \quad (2.40)$$

This enables us to rewrite (2.39) as

$$\psi_{dep} + \gamma_p\sqrt{\psi_{dep}} - (V_{GB} - \phi_{MS} - \psi_s) = 0 \quad (2.41)$$

Assuming ψ_s is known, the above equation can be solved to give

$$\psi_{dep} = \left(-\frac{\gamma_p}{2} + \frac{\gamma_p}{2}\sqrt{1 + \frac{4}{\gamma_p^2}(V_{GB} - \phi_{MS} - \psi_s)}\right)^2 \quad (2.42)$$

To get simpler analytic expressions, (2.42) can be further simplified by noting that for practical values of process parameters and reasonable bias voltages, $\frac{\gamma_p^2}{4} \gg V_{GB}$. Hence, (2.42) can be simplified by using the first two terms of a series expansion of the square root, to give

$$\psi_{dep} = \frac{1}{\gamma_p^2}(V_{GB} - \phi_{MS} - \psi_s)^2 \quad (2.43)$$

In order to compute ψ_s, (2.42) (or the simplified form (2.43)) should be used for ψ_{dep} in (2.36). The resulting equation is implicit and even more complicated that (2.23). Since $\frac{\gamma_p^2}{4} \gg V_{GB}$, ψ_{dep} is a very small fraction of V_{GB}. Hence, we can neglect ψ_{dep} on the right hand side of (2.36), which can now be written as

$$V_{GB} = V_{FB} + \psi_s + \gamma \sqrt{\phi_t \exp(\frac{\psi_s}{\phi_t}) - \psi_s - \phi_t} \qquad (2.44)$$

This is exactly (2.23). Hence, as in the case of an infinitely highly doped gate

$$\psi_s = 2\phi_t \left[\frac{V_{GB} - V_{FB} + 3\phi_t}{V_{GB} - V_{FB} + 6\phi_t} \right] \log\left(1 + \frac{V_{GB} - V_{FB}}{\gamma \sqrt{\phi_t}}\right) \qquad (2.45)$$

If V_{GB} is increased by ΔV_{GB}, by potential balance,

$$\Delta V_{GB} = \Delta \psi_{ox} + \Delta \psi_s + \Delta \psi_{dep} \qquad (2.46)$$

Dividing by the charge per unit area $\Delta Q'_G$ which flows into the gate as a result of this voltage increase,

$$\frac{\Delta V_{GB}}{\Delta Q'_G} = \frac{\Delta \psi_{ox}}{\Delta Q'_G} + \frac{\Delta \psi_s}{\Delta Q'_G} + \frac{\Delta \psi_{dep}}{\Delta Q'_G} = \frac{\Delta \psi_{ox}}{\Delta Q'_G} + \frac{\Delta \psi_s}{-\Delta Q'_c} + \frac{\Delta \psi_{dep}}{\Delta Q'_G} \qquad (2.47)$$

or

$$\frac{1}{C'_{gb}} = \frac{1}{C'_{ox}} + \frac{1}{C'_c} + \frac{1}{C'_{dep}} \qquad (2.48)$$

C'_c is given by (2.29). C'_{dep} is the capacitance of the gate depletion region per unit area. Differentiating (2.37), we get

$$C'_{dep} = \frac{\sqrt{2q\epsilon_s N_{POLY}}}{2\sqrt{\psi_{dep}}} = \frac{\gamma_p C'_{ox}}{2\sqrt{\psi_{dep}}} \qquad (2.49)$$

Using (2.43) in the above equation gives

$$C'_{dep} = \frac{\gamma_p^2 C'_{ox}}{2(V_{GB} - \phi_{MS} - \psi_s)} \qquad (2.50)$$

Hence, the capacitance of the two terminal structure is calculated by using the following four steps

- Step 1: Use the (approximate) explicit relation (2.45) to find ψ_s.

- Step 2: Use the value of ψ_s obtained in Step 1 in (2.29) to calculate C'_c.

- Step 3: Use (2.50) to calculate C'_{dep}.

- Step 4: Use C'_c calculated in Step 2 and C'_{dep} calculated in Step 4 in (2.48) to obtain C'_{gb}.

Notice that C'_{dep} decreases with increasing bias, contrary to C'_c. Thus, there exists the possibility that at some bias voltage, the two effects cancel, and result in a "flat" C–V characteristic. This is precisely what one observes in simulations and measurements. Figure B.1 (Appendix B) shows what happens when gate doping concentration is varied – as N_{POLY} is progressively reduced, the flat top shifts to lower and lower voltages. It can be shown (Appendix B) that an approximate value of V_{GB} at which this happens is

$$V_{GB0} \approx V_{FB} + \gamma_p \sqrt{\phi_t} \tag{2.51}$$

8. MEASUREMENT RESULTS AND DISCUSSION

Figures 2.13 and 2.14 compare our model with measurement results made in two modern n-well salicided CMOS processes. The gate oxide thickness for the data of Figure 2.13 is about 85 Å, while that for Figure 2.14 is 48 Å.

Notice the peak in the capacitor value at about a bias voltage of 2.5 V in Figure 2.13 and around 1.7 V in Figure 2.14. If V_{GB} is increased beyond this point, the capacitance *decreases*. This is the consequence of the polysilicon gate depletion effect, as explained in the previous section. Also, observe that the capacitance peaks at a lower voltage for the 45 Å capacitor. This behavior could have been anticipated from (2.51). As C'_{ox} increases, γ_p decreases, and V_{GB0} is pushed to a lower value.

When oxide thicknesses are further reduced, it has been shown that the charge-sheet model is not quite accurate. The thickness of the charge sheet and quantum mechanical effects need to be taken into account [19] [20].

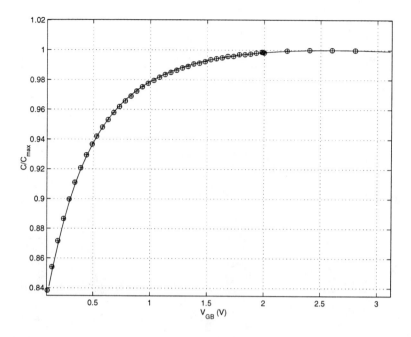

Figure 2.13. Comparison of complete model with measurements. $t_{ox} = 85$ Å(– model, ⊕ data).

9. HIGH FREQUENCY CHARACTERISTICS

The resistance of the channel and the gate are distributed phenomena and cause losses which limit the *Quality Factor* (Q) of the capacitor. This section is devoted to the modeling of these effects. The layout and a three dimensional diagram of a MOS capacitor are shown in Figures 2.15(a) and (b) respectively. The width of the capacitor is W and the length is L. Note that contacts to the n-well and gate are made at *both* ends (metal interconnect is not shown for clarity). The terminals of the device are denoted by A and B. The admittance seen between A and B is denoted as $Y_{in}(s)$. In this analysis, we neglect currents in the v-direction in the n-well and currents in the u-direction in the gate polysilicon. This is justified only if the gate resistivity is much smaller than the well resistivity, which is more than satisfied in typical cases.

Consider the three dimensional diagram of the MOS structure shown in Figure 2.15(b). To analyze the capacitor, it is divided

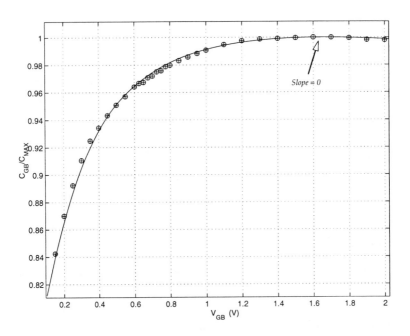

Figure 2.14. Comparison of complete model with measurements. $t_{ox} = 45$ Å (– model, ⊕ data).

into many sections in the v-direction. One such section is shaded. The equivalent circuit for this section is shown in Figure 2.15c. dR_g is the infinitesimal resistance of the gate. dY_c is the infinitesimal admittance of the section from the gate to the terminal B. Given this, the equivalent circuit for the entire capacitor can be drawn as shown in Figure 2.16a. This is redrawn for clarity in Figure 2.16b.

Let R_{ch} be the total channel resistance seen in the u-direction and R_g be the total gate resistance seen in the v-direction. This is depicted in Figure 2.17. Further, let the total capacitance be C. From Figure 2.16, it is clear that, in the v-direction, the equivalent circuit can be drawn as shown in Figure 2.16b. Associating this with Figure C.1 and (C.1) in Appendix C, the admittance of the MOS capacitor is

$$Y_{in}(s) = \frac{4}{R_g}\sqrt{\frac{R_g Y_c(s)}{4}} \tanh\left(\sqrt{\frac{R_g Y_c(s)}{4}}\right) \qquad (2.52)$$

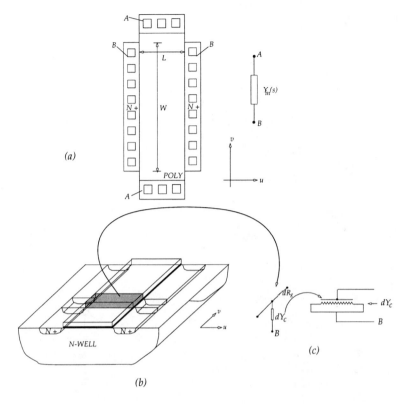

Figure 2.15. Structure of the MOS capacitor and derivation of a two dimensional $\overline{URY}$ equivalent circuit.

Here, $Y_c(s)$ is the admittance when R_g is zero (refer to the discussion of Appendix C). Then, the gate becomes an equipotential surface, and the problem of finding $Y_c(s)$ reduces to a one dimensional $\overline{URY}$ problem. The equivalent distributed network for $Y_c(s)$ is shown in Figure 2.18. Hence, we can use R_{ch} and sC for R and $Y(s)$ in (C.1) to obtain

$$Y_c(s) = \frac{4}{R_{ch}} \sqrt{\frac{sCR_{ch}}{4}} \tanh\left(\sqrt{\frac{sCR_{ch}}{4}}\right) \qquad (2.53)$$

Substitution of $Y_c(s)$ obtained in (2.53) into (2.52) yields

$$Y_{in}(s) = \frac{4}{R_g} \sqrt{\frac{R_g}{R_{ch}} \sqrt{s\tau} \tanh(\sqrt{s\tau})} \tanh\left(\sqrt{\frac{R_g}{R_{ch}} \sqrt{s\tau} \tanh(\sqrt{s\tau})}\right) \qquad (2.54)$$

26 HIGH FREQUENCY CONTINUOUS TIME FILTERS

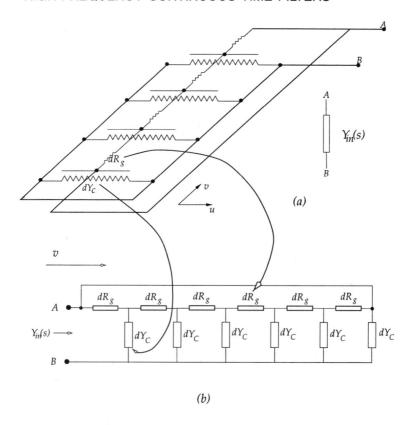

Figure 2.16. Distributed Model for the MOS capacitor.

where

$$R_{ch} = R_{sh}(L/W) \tag{2.55}$$
$$R_g = R_{poly}(W/L) \tag{2.56}$$
$$\tau = R_{ch}C/4 \tag{2.57}$$

R_{sh} and R_{poly} are the sheet resistances of the channel and gate polysilicon respectively.

If the admittance derived in (2.54) is inverted, the driving point impedance of the capacitor is obtained. Expanding the hyperbolic tangents of the impedance function in a Taylor series and keeping the first two terms, we get

$$Z_{in}(s) \approx \frac{1}{Y_{in}(s)} = \frac{1}{sC} + \frac{(R_{sh}(L/W) + R_{poly}(W/L))}{12} \tag{2.58}$$

MOS Capacitor Modeling

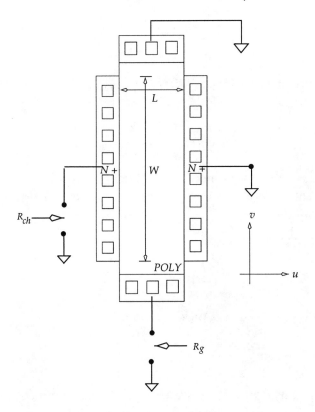

Figure 2.17. Definitions of R_{ch} and R_g.

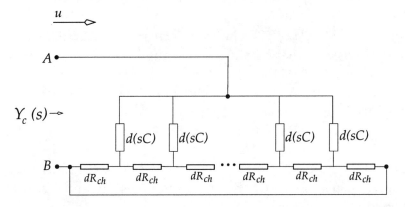

Figure 2.18. Distributed network to calculate $Y_c(s)$.

Hence, the "equivalent impedance" of the MOS capacitor for low frequencies (that is where $\frac{R_g}{R_{ch}}\omega\tau \ll 1$) is an $R-C$ series circuit. The capacitance is C and the series resistance is

$$R_{series} = \frac{(R_{sh}(L/W) + R_{poly}(W/L))}{12} \qquad (2.59)$$

The quality factor, $Im(Z_{in}(j\omega))/Re(Z_{in}(j\omega))$, of the capacitor can now be written as

$$Q \approx \frac{12}{\omega C'_{ox}(R_{sh}L^2 + R_{poly}W^2)} \qquad (2.60)$$

where C'_{ox} is the capacitance of the oxide per unit area. For a given capacitor value, WL is fixed. There thus exists an optimum aspect ratio for the capacitor which gives the highest quality factor, and this is given by

$$\left(\frac{W}{L}\right)_{optimum} = \sqrt{\frac{R_{sh}}{R_{poly}}} \qquad (2.61)$$

Under this optimal condition,

$$Q_{max} = \frac{6}{\omega C'_{ox} R_{sh} L^2} \qquad (2.62)$$

Note that this optimum assumes that the interconnect and contacts have no resistance. For a given process, the optimal aspect ratio can be rederived using the contact resistances in series with $Z_{in}(s)$ in (2.58).

To verify (2.54), a SPICE circuit was created with grids of resistors and transmission lines as shown in Figure 2.19. The capacitance of the structure was 0.75 pF. R_g and R_{ch} were 30Ω and 50Ω respectively. The real and imaginary parts of the admittance of the capacitor are plotted in Figures 2.20 and 2.21 respectively. In these figures, "simple model" stands for the model of (2.58). For the particular values of R_{ch} and C, $\tau = 9.37 \times 10^{-12}$s.

From the figures, it is clear that the simple model is fairly accurate at low frequencies. Also, (2.54) agrees very well with the admittance obtained by simulation.

10. N-WELL TO SUBSTRATE CAPACITANCE

The bottom plate of the capacitor is the *n*-well. There is a reverse biased junction between the *n*-well and the substrate. Since the substrate is grounded, the depletion region can be modeled as a

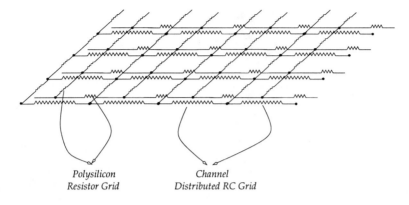

Figure 2.19. Part of a grid to simulate 2-D effects in SPICE.

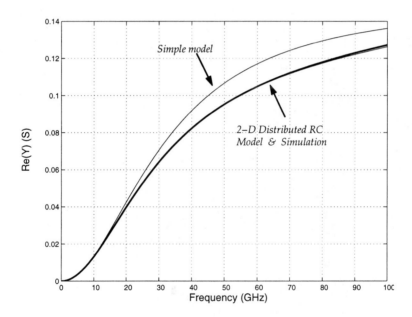

Figure 2.20. Re(Y) computed using various models.

capacitance between the n-well and ground. The usual formulae from p-n junction theory can be used to determine the value of this capacitance. As technologies scale, gate oxide thicknesses reduce and doping concentrations increase. Hence, the ratio of the parasitic depletion capacitance to the gate capacitance cannot be expected to change dramatically. For example, in the technology used for test

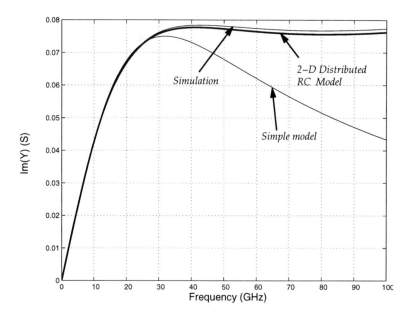

Figure 2.21. Im(Y) computed using various models.

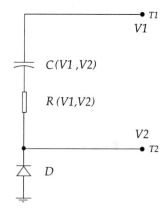

Figure 2.22. Complete model for the poly-n-well MOS accumulation capacitor.

chips in this book, this ratio was about 30%. The final model for the capacitor, including all parasitics is shown in Figure 2.22.

11. SUMMARY

The use of MOS accumulation capacitors is not only feasible but also desirable for many kinds of filter specifications. Their attractive properties are

- High specific capacitance

- Good control of absolute capacitor value

- Excellent capacitor matching

- Compatibility with a digital CMOS process

Whether the distortion levels they produce are acceptable depends on the situation at hand. Figures 2.23 and 2.24 show the second and third harmonic distortion components of capacitor voltage when driven by a sinusoidal current. The figures are for the case when there is no gate depletion, and hence are a conservative estimate. Note that even for a bias voltage of 1 Volt, third harmonic distortion is less than 60 dB for a 0.25 V peak signal. This corresponds to a 1 V peak-to-peak signal when two capacitors are operated in a fully differential fashion. Second harmonic distortion is about 50 dB under the same conditions. The performance of circuits that use these capacitors will be high even in the presence of mismatches because even harmonic components are inherently low.

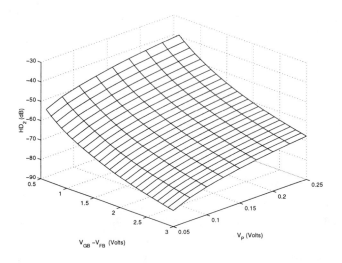

Figure 2.23. HD_2 as a function of bias voltage and signal level.

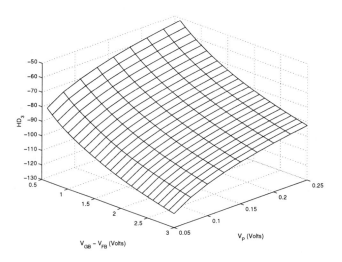

Figure 2.24. HD_3 as a function of bias voltage and signal level.

Chapter 3

A REVIEW OF INTEGRATOR ARCHITECTURES

1. INTRODUCTION

In this chapter, we compare several integrator structures commonly used in the design of monolithic filters. The integrator is the basic building block of these filters, hence the performance of various architectures can be evaluated by studying the properties of the corresponding integrator structures. The next section is devoted to the analysis of non-ideal effects in integrators and their effect on the frequency response of a biquadratic section. Then, we examine various circuit design issues and tradeoffs involved in three popular filter techniques. CMOS transconductors are compared based on several performance parameters. In the last section of this chapter, we discuss the additional problems that arise when filters have to be programmed over a wide range.

2. NON-IDEALITIES IN INTEGRATORS AND THEIR EFFECT ON BIQUADRATIC SECTIONS

The magnitude and phase responses of an ideal integrator are shown in dotted lines in Figure 3.1. In practice, the response deviates from the ideal at low frequencies due to finite DC gain and at high frequencies due to parasitic poles and zeros. One possible situation is shown by the solid line in Figure 3.1, where a high frequency parasitic pole is assumed.

A convenient method of evaluating integrator performance is in terms of its quality factor, $Q_I(\omega)$. The quality factor of an integrator

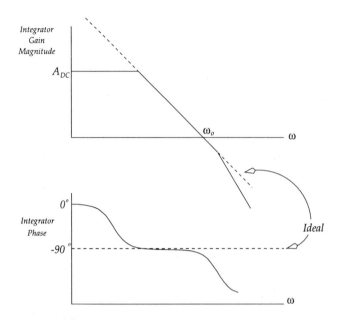

Figure 3.1. Magnitude and phase responses of an integrator

is analogous to that of an inductor or capacitor and is a measure of the integrator's phase deviation from the ideal $-90°$. If the nonideal integrator frequency response is placed in the form of

$$H_I(j\omega) = \frac{1}{R(\omega) + jX(\omega)} \tag{3.1}$$

then the quality factor is defined as

$$Q_I(\omega) = \frac{X(\omega)}{R(\omega)} \tag{3.2}$$

Note that ideally,

$$H_I(j\omega) = \frac{1}{j\left(\frac{\omega}{\omega_o}\right)} \tag{3.3}$$

and

$$Q_I(\omega) = \infty \tag{3.4}$$

We now examine an ideal biquad and then consider the effect of integrator phase and gain errors on biquad performance [37]. The analysis is general and applicable to biquads constructed with arbitrary integrators. The pole-forming section of a double integrator

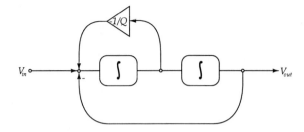

Figure 3.2. Pole forming section of a biquad

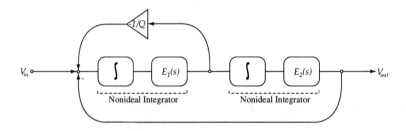

Figure 3.3. Pole forming section of a biquad with non-ideal integrators

loop is shown in Figure 3.2. Both integrators are identical with transfer functions $\frac{\omega_o}{s}$. The biquad loop gain is the product of the individual transfer functions around the outermost loop encompassing the damped and undamped integrators. The denominator polynomial of any transfer function realized is

$$D(s) = \frac{s^2}{\omega_o^2} + \frac{s}{\omega_o Q} + 1 \tag{3.5}$$

When this is evaluated on the $j\omega$ axis, we get

$$D(j\omega) = \left(1 - \frac{\omega^2}{\omega_o^2}\right) + j\frac{\omega}{\omega_o Q} \tag{3.6}$$

The real part of the denominator is zero when $\omega = \omega_o$. At this frequency, the imaginary part is $1/Q$. Turning things upside down, we now say that the *quality factor* of the pole forming section is the reciprocal of the imaginary part of the denominator polynomial (also called the characteristic polynomial) at that frequency at which the real part is zero. A biquadratic loop with non-ideal integrators is shown in Figure 3.3. Let us assume that $E_1(s)$ and $E_2(s)$ model parasitic poles in the integrator - that is,

$$E_1(s) = 1/(1 + \frac{s}{\omega_1}) \tag{3.7}$$

$$E_2(s) = 1/(1 + \frac{s}{\omega_2}) \tag{3.8}$$

The denominator polynomial of the system now becomes

$$D'(s) = \frac{s^2}{\omega_0^2}\left(1 + \frac{s}{\omega_1}\right)\left(1 + \frac{s}{\omega_2}\right) + \frac{s}{\omega_0}\left(\frac{1}{Q} + \frac{s}{Q\omega_2}\right) + 1 \tag{3.9}$$

Expanding the above, we get

$$D'(s) = \frac{s^4}{\omega_0^2 \omega_1 \omega_2} + \frac{s^3}{\omega_0^2}\left(\frac{1}{\omega_1} + \frac{1}{\omega_2}\right) + \frac{s^2}{\omega_0^2}\left(1 + \frac{\omega_0}{Q\omega_2}\right) + \frac{s}{Q\omega_0} + 1 \tag{3.10}$$

If the parasitic poles are far removed from ω_0, an approximate value for ω at which the real part of $D'(j\omega)$ goes to zero is ω_0. At this frequency,

$$D'(j\omega_0) \approx j\left(\frac{1}{Q} - \frac{\omega_0}{\omega_1} - \frac{\omega_0}{\omega_2}\right) \tag{3.11}$$

From our definition of biquad quality factor, we can express Q', the quality factor of the pole forming section built using nonideal integrators as

$$\frac{1}{Q'} = \left(\frac{1}{Q} - \frac{\omega_0}{\omega_1} - \frac{\omega_0}{\omega_2}\right) \tag{3.12}$$

Recalling the definition of integrator quality factor (3.2), and using (3.7) and (3.8), we observe that

$$Q_{I1}(\omega_0) \approx -\frac{\omega_1}{\omega_0} \tag{3.13}$$

$$Q_{I2}(\omega_0) \approx -\frac{\omega_2}{\omega_0} \tag{3.14}$$

and (3.12) can be rewritten as

$$\frac{1}{Q'} = \frac{1}{Q} + \frac{1}{Q_{I1}} + \frac{1}{Q_{I2}} \tag{3.15}$$

If both integrators are identical, (3.15) reduces to

$$\frac{1}{Q'} = \frac{1}{Q} + \frac{2}{Q_I} \tag{3.16}$$

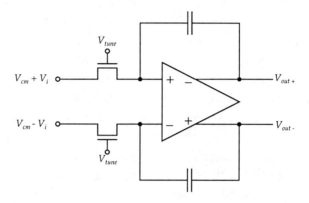

Figure 3.4. A MOSFET-C integrator

The effect of nonideal integrators is to modify the quality factor of the biquad[43]. For example, if a biquad with a quality factor of 50 has to be implemented with a Q accuracy of 1%, we would need an integrator quality factor of 10000!

Depending on the implementation of the integrators, most filter types fall under one of the three following categories :

- MOSFET-C
- Gm-C
- Gm-OTA-C

We consider these three architectures in detail [22] [23], and compare them on the basis of various criteria - such as high frequency capabilities, demands made on the active devices used, sensitivity to parasitics, power consumption and most importantly, the capability of being used with MOS capacitors. The aim of this exercise is to be able to weed out architectures which certainly cannot meet our requirements.

3. MOSFET-C FILTERS

The MOSFET-C technique [24] [25] [26] [27] dates back to the early days of CMOS filters. A schematic of the MOSFET-C integrator is shown in Figure 3.4. Basically, this is a reincarnation of the classical active RC integrator in a fully balanced form. But unlike its classical ancestor, fully balanced operation is essential to the operation of the MOSFET-C integrator. The MOSFETs operate in their linear or triode region - to guarantee that this is so, V_{tune} must be sufficiently high. The even order non-linearities cancel out due to balanced operation and the odd order non-linearities are sufficiently small for many applications. The active element can be either an operational amplifier or an operational transconductance amplifier. We will consider each of these cases separately. When the active device is an opamp, the integrator is insensitive to parasitic capacitance at the output nodes due to the opamp's low output impedance. If the filter corner frequency is very high, then the impedance levels in the filter are low, and the opamp has to have a much smaller output impedance. The low output impedance of the opamp also eliminates the feed-forward effect of the integrating capacitance which would otherwise give rise to a right half plane zero in the integrator transfer function. Having a low impedance output stage means increased power dissipation, especially in CMOS technologies. On the other hand, if an OTA is used as the active element its transconductance must be a few orders of magnitude larger than the conductance of the MOSFET [4] [22]. Use of the OTA causes a right half plane zero in the integrator transfer function. Techniques do exist to eliminate or cancel the effect of the right half plane zero, but these are difficult to apply reliably. Phase compensation strategies have been proposed to reduce the effect of finite bandwidth of the opamp or OTA. These increase complexity and their efficacy has not been proved to date in high frequency filters.

A disadvantage of the MOSFET-C integrator in the context of digital VLSI processes is that it uses floating capacitors. This is certainly not convenient if accumulation capacitors have to be used, since the use of the MOSFET fixes the input and output quiescent voltages of the opamps or OTAs in the filter to be the same. Therefore, the DC voltage across the capacitor is constrained to be zero. The integrator shown in Figure 3.4 cannot be used directly with MOS capacitors. Alternatives which might work are now considered and the various trade offs are discussed.

Figure 3.5 shows schematics of four possible modifications of the traditional MOSFET-C integrator enabling it to be used with MOS accumulation capacitors. From the considerations in the previous

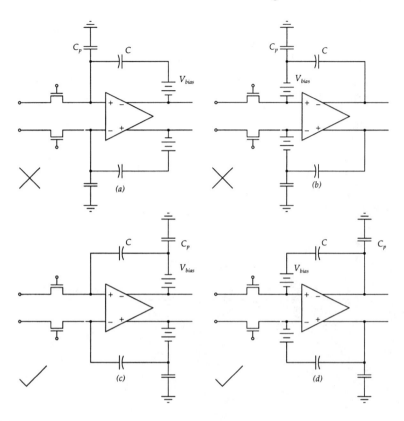

Figure 3.5. Conceptual schematics of MOSFET-C integrators with accumulation capacitors.

chapter, the value of V_{bias} is chosen to be 1 volt. Of the four schematics shown, the ones of Figure 3.5(a),(b) are to be avoided because C_p can be as large as one third of C. This means that there is a low impedance path from the substrate to the input of the opamp, and this could result in significant amounts of substrate noise coupling into the filter. The circuits of Figure 3.5(c),(d) are preferable because the parasitic is driven by the output of the opamp which is a low impedance node. The DC level shifter, shown as a battery of value V_{bias} can be implemented in several ways. It becomes convenient to combine the level-shifting with either a voltage or current buffer, as will be seen.

Keeping in mind that the feedback around the opamp is of the voltage-shunt kind, and substituting various implementations of the battery into the schematics shown in Figure 3.5(c)&(d) we get the

40 HIGH FREQUENCY CONTINUOUS TIME FILTERS

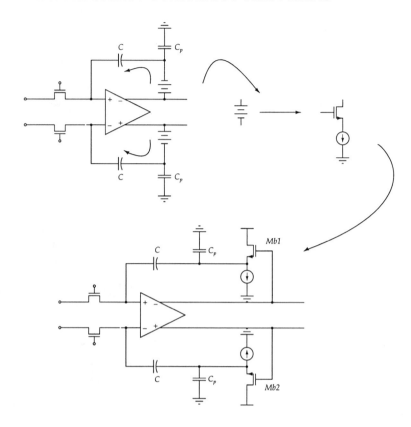

Figure 3.6. A MOSFET-C integrator using accumulation capacitors.

integrator topologies shown in Figure 3.6 and 3.7. We analyze each of these in some detail. The use of an *n*-well technology is assumed here, so that the substrate terminal of an NMOS device is grounded while that of a PMOS device can be connected to its source.

Consider the integrator of Figure 3.6. Mb1 and Mb2 act as voltage buffers and at the same time provide the required bias voltage across the MOS capacitor. Their gate-source and body-source transconductances are denoted below by g_m and g_{mb} respectively. For the time being, we assume that the operational amplifier is ideal. The transfer function of the integrator can be shown to be

$$H(s) = -\left\{\frac{1}{sCR}\left(\frac{g_m+g_{mb}}{g_m}\right)\right\}\left(1+\frac{s(C+C_p)}{g_m+g_{mb}}\right) \qquad (3.17)$$

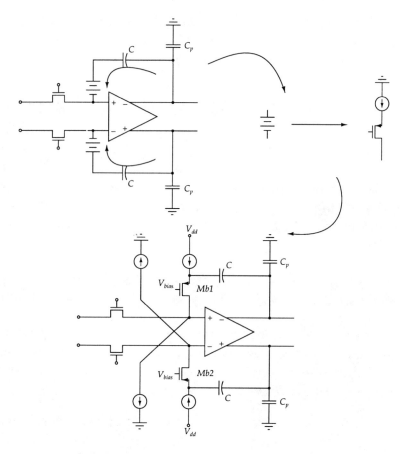

Figure 3.7. An alternate MOSFET-C integrator using accumulation capacitors.

The above equation can be rewritten as

$$H(s) = -\frac{\omega_o}{s}(1 + \frac{s}{\omega_z}) \qquad (3.18)$$

where

$$\omega_o = \frac{1}{RC}\left(\frac{g_m + g_{mb}}{g_m}\right) \qquad (3.19)$$

and

$$\omega_z = \frac{g_m + g_{mb}}{(C + C_p)} \qquad (3.20)$$

The buffer modifies the value of the integrator unity gain frequency by a factor $\frac{g_m}{g_m+g_{mb}}$, and causes a left half plane zero to appear at ω_z. ω_z will have to be higher than ω_o by at least an order of magnitude. This means that the buffer will have to draw a large bias current.

42   HIGH FREQUENCY CONTINUOUS TIME FILTERS

Another design is shown in Figure 3.7. Here, the current through the capacitance is buffered. Under the same assumptions as in the voltage buffer case, the transfer function of the "integrator" can be shown to be

$$H(s) = -\left\{\frac{1}{sCR}\right\}\left(1 + \frac{sC}{g_m}\right) \qquad (3.21)$$

We still have not got rid of the left half plane zero. However, ω_o is equal to the desired value of $\frac{1}{RC}$ and ω_z does not depend on C_p.

Hence, one big advantage of the MOSFET-C structure is lost if one tries to utilize a floating capacitor. The effects of finite opamp gain and bandwidth on the transfer function of the integrator are straightforward, although algebraically tedious. We will not pursue these calculations.

We now summarize the properties of the MOSFET-C integrator listing its advantages and disadvantages:

- (+) It can be very linear. Its tuning range is only limited by the supply voltage. By using charge pump techniques, voltages greater than the supply can easily be realized. Alternatively "range switching" can be resorted to.

- (-) The operational amplifier has to drive resistive loads; however, a large transconductance OTA can be used instead. This would make sense mostly in BiCMOS technologies [4] [22].

- (-) Operation with accumulation capacitors is not directly possible. The modifications suggested in Figures 3.6 and 3.7 can be used but they do not have the low sensitivity properties of the basic MOSFET-C structure - apart from being power hungry.

Of the several disadvantages we have discussed above, the most significant one is the difficulty of using accumulation capacitors.

4.   GM-C FILTERS

The Gm-C filter technique [2] [8] [28] [30] [54] is based on an open-loop integrator structure. Two versions of the Gm-C integrator are shown in Figure 3.8. In both cases, a transconductor produces a current proportional to the differential input voltage and the output is taken across the integrating capacitors. Figure 3.8(a) shows an

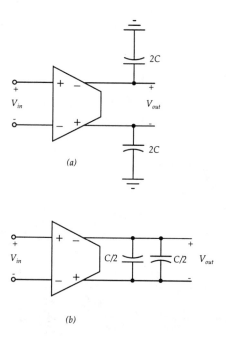

Figure 3.8. Two versions of the Gm-C integrator.

integrator using grounded capacitors, while (b) employs a floating capacitor. If the transconductor is ideal with a transconductance equal to G_m, the transfer function of both integrators can be shown to be

$$H(s) = \frac{G_m}{sC} \qquad (3.22)$$

Note that the integrator of Figure 3.8(a) uses four times the capacitance of that used in Figure 3.8(b).

The transconductor is a voltage controlled current source required to have infinite input and output impedances, linearly handle large signals and possess a very stable transconductance since the unity gain frequency of the integrator is proportional to it. Since the circuit in Figure 3.8 is an open-loop integrator, it has the potential for very high speeds. A significant departure from the MOSFET-C integrator is that the transconductor sees only a capacitive load. The Gm-C integrator is in some sense a natural approach to active filter design because all active devices (BJT, MOSFET) are inherently transconductance elements. The DC gain, linearity and the high frequency performance of the integrator is heavily dependent on the

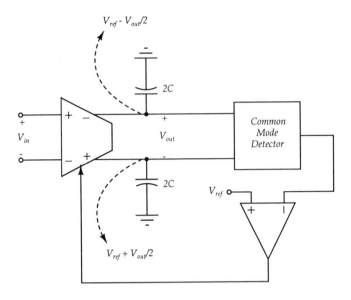

Figure 3.9. The Gm-C integrator with the common-mode setting loop explicitly shown.

particular transconductor used in the design. A discussion of the merits and demerits of particular transconductor designs here can only obfuscate our understanding of the topological properties of Gm-C filters. It will therefore be postponed for a later occasion. An important advantage of using the grounded capacitor version of the Gm-C integrator (Figure 3.8(a)) is that it is ideally suited for use with non-linear capacitors. This is shown in Figure 3.9. As in any fully differential system, there is a common-mode feedback circuit which holds the common-mode level (or the quiescent voltage level when the differential input signal is zero). The CMFB (Common-mode feedback) loop forces the output common mode voltage of the transconductor to be V_{ref}, irrespective of process variations or temperature. This means that the non-linear capacitor can be made to have a fixed bias voltage across it. The capacitor and the CMFB loop have a symbiotic relationship - as far as the CMFB circuit is concerned, the grounded capacitor serves to stabilize the loop which might otherwise become unstable because of the high gains involved. For the particular case of MOS accumulation capacitors, the parasitic n-well to substrate depletion capacitance is not of concern anymore as the bottom plate is grounded. The application of more than a volt of bias across the capacitor is easily achieved. The fact that we use four times more capacitance than the integrator of Figure 3.8(b) is not a problem - the specific capacitance of the MOS capacitors is high, and the area occupied by the capacitors is small.

The only disadvantage of the Gm-C integrator is its sensitivity to parasitic capacitances, which cause a change in integrator unity gain frequency. Interconnect and depletion parasitics have to be carefully modeled and included as a part of the filter design. This is problematic because the parasitics might not be accurately known. In very high frequency designs, the parasitics might be a significant part of the integrating capacitance - poorly characterized parasitics can cause a deterioration in the matching of capacitor ratios. More importantly, the parasitics may not track the integrating capacitors in the presence of process and temperature variations. This is not as bad as it seems at first - even the MOSFET-C integrator, when used with MOS capacitors has a transfer function dependent on the depletion capacitor between the well and the substrate (if the output impedance of the opamp is not small). For high performance designs, parasitics have to be characterized well anyway and the filter will require careful device level design and layout.

We now summarize the properties of the Gm-C technique, listing its advantages and disadvantages

- (+) The transconductor has to drive only capacitive loads.

- (+) Operation with accumulation capacitors is trivial - the area overhead due to the use of grounded capacitors is not a real issue because of the high density of MOS capacitors.

- (-) The integrator unity gain frequency is sensitive to parasitic capacitances.

Of the two advantages we have discussed above, the most significant one is the ability to use MOS accumulation capacitors as the integrating elements. Thus, the Gm-C technique seems to be attractive for realizing very high frequency filters in a standard CMOS process.

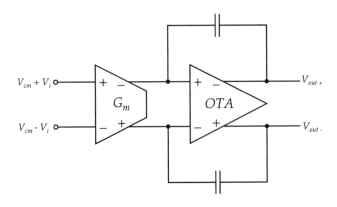

Figure 3.10. A conceptual schematic of a Gm-OTA-C integrator.

5. GM-OTA-C FILTERS

The Gm-OTA-C integrator [3] [7] combines some of the advantages of the MOSFET-C and Gm-C structures when conventional capacitors are used. The conceptual schematic of the Gm-OTA-C integrator is shown in Figure 3.10. It can be derived from the MOSFET-C integrator as follows. The opamp is replaced by an operational transconductance amplifier (OTA), which is required to have a very large transconductance and a high output impedance. The MOSFETs are replaced by a transconductor which is required to be linear for the expected signal swings in the filter. The outputs of the transconductor, however are the input terminals of the OTA and hence are at a constant potential. The parasitic capacitance at the transconductor output sees little voltage excursion and is hence immaterial. Moreover, the OTA is enclosed in a unity gain feedback loop, and hence the integrator may have a higher frequency capability compared with the MOSFET-C case. The integrator's low frequency gain is the product of the gains of the OTA and the transconductor and can be made relatively large. This could be a very useful property in CMOS implementations because short channel MOSFETs (required for high speed operation) have low output impedances and implementing high gains with them is difficult. As in the MOSFET-C case, operation with a floating capacitor presents problems. We go through the same chain of thought as done in the MOSFET-C case to derive integrator topologies using accumulation capacitors. Figure 3.11 shows a scheme which uses a source follower

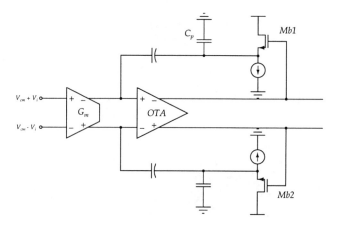

Figure 3.11. Schematic of a Gm-OTA-C integrator with accumulation capacitors.

as a level shifter. The structure of Figure 3.12 uses a current buffer to perform the level shift function. Figure 3.13 shows another integrator arrangement where the transconductor current is injected at the input of the current buffer, as opposed to the virtual ground of the OTA. After some algebra it can be shown that this integrator has two zeros, symmetrically placed about the $j\omega$ axis. This works quite well in the case where the current buffer has a very low input impedance (for example, when bipolar transistors are used in lieu of Mb1 and Mb2) [29]. However, a CMOS implementation would necessitate large bias currents in Mb1 and Mb2.

As was discussed in the MOSFET-C case both these schemes give away the low parasitic sensitivity properties of the original integrator structure. The Gm-OTA-C integrator, when implemented with accumulation capacitors is sensitive to the parasitic bottom-plate capacitor. Moreover, the high transconductance demanded of the OTA will necessarily mean a considerable input capacitance (not to mention power, area and layout complexity). In the authors opinion, it is not easy to use the Gm-OTA-C technique in a digital CMOS process. The DC gain of the integrator being high is only a small consolation in a morass of serious problems.

After our study of various filter architectures, we conclude that the Gm-C technique is best suited for CMOS filters in the very high frequency range built in a digital CMOS process. Our task now is to carefully study the problems of a Gm-C architecture. Since

48   HIGH FREQUENCY CONTINUOUS TIME FILTERS

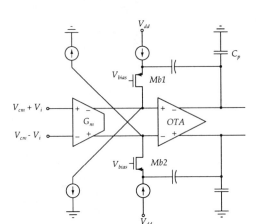

Figure 3.12. Alternative schematic of a Gm-OTA-C integrator with accumulation capacitors.

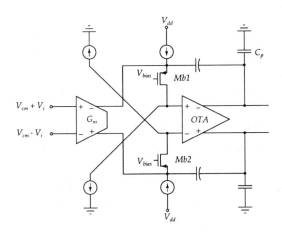

Figure 3.13. Another Gm-OTA-C integrator with accumulation capacitors.

the properties of a Gm-C integrator are heavily dependent on the particular transconductor in question, in the next section we take it upon ourselves to converge on possible transconductor topologies.

6.   A STUDY OF CMOS TRANSCONDUCTORS

CMOS transconductors have been extensively studied over the years. Three candidate transconductor families will be compared.

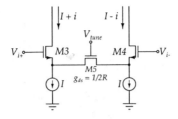

Figure 3.14. A degenerated differential pair.

6.1 TRANSCONDUCTORS BASED ON THE DEGENERATED DIFFERENTIAL PAIR

The simplest circuit of this family [31] is shown in Figure 3.14. The differential pair formed by M3-M4 is degenerated by a MOSFET operating in the triode region. The resistance of the MOSFET is denoted by $2R$ and is tunable by V_{tune}. It can be shown that (neglecting the body effect of M3 and M4)

$$G_m = \left(\frac{i}{v_{i+} - v_{i-}}\right) = \frac{1}{2R}\left(\frac{g_{m3}R}{g_{m3}R + 1}\right) \qquad (3.23)$$

where

$$2R = \frac{1}{\mu_n C'_{ox}\left(\frac{W}{L}\right)(V_{GS,M5} - V_T)} \qquad (3.24)$$

In using this arrangement, it is understood that $g_{m3}R$ should be made large - otherwise it is not possible to tune the transconductor. The design of this type of circuit presents problems in low voltage CMOS technologies, and especially in our case where large transconductance values are desired. A schematic showing all parasitic effects and excess noise sources is shown in Figure 3.15. The transconductor has internal nodes - parasitic depletion capacitances (which can be fairly large) at the source nodes of M3 and M4 cause phase shifts at high frequencies. Ideally, the input impedance of the transconductor should be very large. This is very difficult to achieve when the desired G_m is large because of the input capacitance of the large devices that need to be used. The next best choice is that the impedance should be purely capacitive in which case it can be a part of the integrating capacitors in the filter. The differential input impedance of a triode degenerated differential pair looking into the

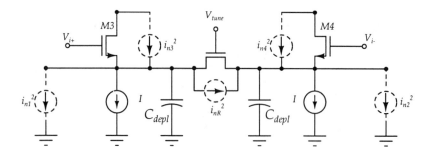

Figure 3.15. A degenerated differential pair with parasitic effects and excess noise sources.

gates of M3 and M4 is a complicated function of frequency - more specifically, it is not a pure capacitance.

The input referred noise voltage spectral density of a transconductor (excluding $1/f$ noise) can be written as

$$\frac{\overline{v_n^2}}{\Delta f} = \frac{4kT}{G_m}\eta \qquad (3.25)$$

where G_m is the transconductance value and η is called the excess noise factor. η is an index of how noisy the transconductor is, since the quantity it multiplies in (3.25) is approximately the lowest noise that can be achieved by a MOS transconductor with transconductance G_m [16]. As thermal and shot noise are often dominant in very high frequency applications, we neglect $1/f$ noise above, and in the rest of this work. A little thought shows that the noise performance of the degenerated differential pair leaves a lot to be desired. Since g_{m3} and g_{m4} are large in relation to $1/R$, a large bias current I will be required to flow through them. The choice of input common-mode voltage (typically around half the supply voltage) and bias current constrain the aspect ratio of the tail current sources to be around the same as that of M3 and M4. This means that they will have a large transconductance. The noise currents due to the tail current sources, which will be many times that generated by M5, appear at the output of the transconductor. If we assume that the devices forming the tail current sources and the input devices M3-M4 all have the same transconductance g_m, it can be shown that the excess noise factor is approximately given by

$$\eta \approx \frac{2}{3} g_m R \qquad (3.26)$$

This calculation does not take into account the noise contributed by the PMOS load devices. One can see that this transconductor tends to be very noisy. Another possible problem is the following. If the input common-mode voltage is changed (presumably due to mismatches or inaccuracies in common-mode feedback loops), the gate overdrive of M5 is altered changing R and thereby G_m. As the tuning voltage is varied, the circuit's linear range changes. Any mismatch in the tail currents will result in transconductor offset. These factors degrade the dynamic range.

In summary, the properties of a degenerated differential pair are :

- (+) Can be very linear

- (-) Has internal nodes causing phase-shifts at high frequencies

- (-) Large excess noise factor

- (-) Input impedance is not purely capacitive

- (-) Sensitive to input common-mode shifts

- (-) Mismatch in tail currents reflects as input offset

In view of all the above, the large linear range and tunability are offset by several disadvantages. We conclude that it is difficult to design very high frequency filters using this transconductor. Figure 3.16 shows a variation proposed in [30]. While this circuit is not sensitive to common-mode inputs, it still retains most of the problems discussed above. Other modifications of the basic transconductor of Figure 3.14 are based on Figure 3.17. It is clear that anything as complicated as this will not be easy to design for very wideband operation.

6.2 TRANSCONDUCTORS BASED ON MOSFETS WITH CONSTANT DRAIN-SOURCE VOLTAGE

The conceptual schematic of this family of circuits [32] is shown in Figure 3.18. M3 and M4 operate in the triode region. The function

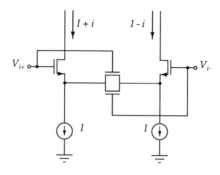

Figure 3.16. The transconductor of Krummenacher and Joehl.

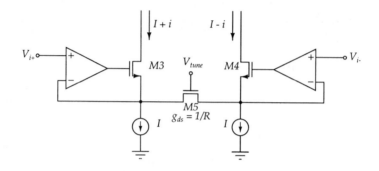

Figure 3.17. Transconductors based on opamps.

of M1, M2 and the associated opamps is to pin the drains of M3 and M4 at a fixed potential V_{tune}. Then,

$$I_1 = \mu_n C'_{ox}\left(\frac{W}{L}\right)\left[(V_{i+} - V_T)V_{tune} - a\frac{V_{tune}^2}{2}\right] \quad (3.27)$$

$$I_2 = \mu_n C'_{ox}\left(\frac{W}{L}\right)\left[(V_{i-} - V_T)V_{tune} - a\frac{V_{tune}^2}{2}\right] \quad (3.28)$$

where a is a process dependent parameter somewhat larger than unity [16]. Hence,

$$I_1 - I_2 = \mu_n C'_{ox}\left(\frac{W}{L}\right)(V_{i+} - V_{i-})V_{tune} \quad (3.29)$$

$$G_m = \mu_n C'_{ox}\left(\frac{W}{L}\right)V_{tune} \quad (3.30)$$

In practice, second-order effects like mobility reduction and velocity

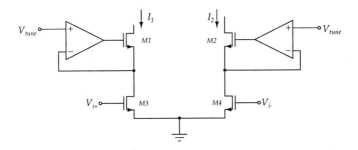

Figure 3.18. Conceptual schematic.

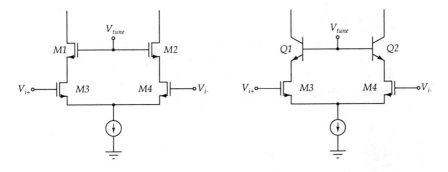

Figure 3.19. Realizations in CMOS and BiCMOS.

saturation reduce the linearity somewhat. In very high frequency filters having full-fledged operational amplifiers is not practical. In CMOS, a source follower is used. In BiCMOS implementations, an emitter follower does the job. Both these circuits are shown in Figure 3.19. The transconductor has internal nodes. These cause high frequency poles in the transconductance realized. It can be shown that the main parasitic pole occurs at the drains of M3 and M4. In a CMOS realization, this means that the g_m's of M1 and M2 (Figure 3.19) must be very high - leading to high power consumption. Maintaining a low impedance at the drains of the MOSFETs is much easier to do in BiCMOS technology - in fact, the BiCMOS version of this transconductor has been the basis of several high frequency filters reported in the literature [33]. The excess noise factor of this circuit is smaller than that of the degenerated differential pair because the noise of the tail current source appears as a common-mode signal. There is a tradeoff between transconductance value and the excess

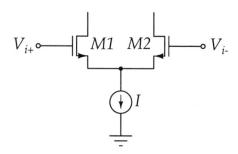

Figure 3.20. The differential pair.

noise factor, which is especially serious in a CMOS implementation. In [33], even a BiCMOS implementation has η ranging from 4-10.

In summary, the properties of the CMOS version of this transconductor are :

- (+) High linearity

- (+) Tunable over a wide range

- (-) High excess noise factor

- (-) Has internal nodes causing phase-shifts at "high" frequencies - this is very problematic in CMOS.

6.3 THE DIFFERENTIAL PAIR

The differential pair is the simplest of all transconductors, with several appealing properties for high frequency work. A schematic is shown in Figure 3.20. When operated in a fully differential fashion, the source coupled node acts as a virtual ground for small signals. Therefore, the transconductor has no internal signal carrying nodes and no parasitic poles. Noise from the tail current source appears only as a common-mode component. A small signal equivalent circuit for the differential pair is shown in Figure 3.21, where a simple quasistatic model has been assumed for the MOSFETs. The excess noise factor can be shown to be

$$\eta \approx 2/3 \tag{3.31}$$

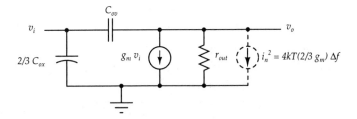

Figure 3.21. Small signal single ended equivalent circuit of a differential pair.

If the MOSFET is assumed to behave like a square law device in saturation according to

$$I_{DS} = \frac{\mu_n C'_{ox}}{2a} \left(\frac{W}{L}\right)(V_{GS} - V_T)^2 \qquad (3.32)$$

the transconductance is

$$g_m = \frac{2 I_{DS}}{V_{GS} - V_T} = \sqrt{\frac{\mu_n C'_{ox} \left(\frac{W}{L}\right)}{a} I} \qquad (3.33)$$

The main distortion component in the differential output current is of the third order and it can be shown that

$$HD_3 = \frac{1}{8}\left(\frac{V_P}{V_{GS} - V_T}\right)^2 \qquad (3.34)$$

where V_P is the peak voltage swing on either input. From (3.34), we see that the linear range of the differential pair can in principle be made arbitrarily large by increasing the "gate overdrive voltage", $V_{GS} - V_T$. If V_P is of the order of 50 mV, a gate overdrive of 250 mV is sufficient for third harmonic output current levels about 46 dB below the fundamental. In practice, there will also be even harmonic distortion due to device mismatches. The gate overdrive will need to be higher than 250 mV in the worst case.

Another important property of the differential pair is that its input impedance is purely capacitive. (In stating this, we have neglected nonquasistatic effects in the MOS transistor and the effect of the gate-drain overlap capacitance, which will give a real part to the input impedance at very high frequencies [16].) In our case, where the integrating capacitance is formed by gate oxide, this feature is very attractive because the "parasitic" transconductor input capacitance and the integrating capacitors are of the same kind, and will thus track. This means that capacitor ratios can be expected to be very stable over process and temperature variations. Another useful feature is that to a first order, the value of g_m realized is fairly insensitive to input common-mode voltage.

The only way to tune the transconductance of the differential pair is to vary the bias current. This is not attractive if the filter time constants have to be tuned over a wide range. For example, if the g_m has to be changed by a factor of 5, one would need to increase bias currents by a factor of 25 (see (3.33))! Therefore, only fine tuning is practical through the tail current. This is not as serious a disadvantage as it seems at first. "Range switching" techniques can be used to effect large changes in transconductor value. In summary, the properties of a differential pair are :

- (+) No internal nodes

- (+) Low excess noise

- (+) Input impedance purely capacitive

- (+) Insensitive to input common-mode shifts

- (-) Only limited tunability is possible

7. PROBLEMS OF PROGRAMMABLE FILTER DESIGN

Apart from the usual issues associated with high frequency CMOS filter design discussed in this chapter, the issue of *programmability* brings to the forefront the very important problem of *maintaining* performance indices like frequency response accuracy, noise and dynamic range across the *entire* tuning range. We will now examine each of these issues. At the end of the discussion, we will have established a set of desirable attributes of a good programmable integrator.

7.1 MAINTAINING FREQUENCY RESPONSE SHAPE IN WIDELY PROGRAMMABLE FILTERS

In order to understand what it takes on the part of an integrator to achieve a frequency response whose relative shape does not change with bandwidth setting, consider the design of a programmable biquad, based on the double integrator structure of Figure 3.2. Let the lowest center frequency setting desired of the biquad be ω_{lowend}, and the quality factor needed be $Q_{desired}$. Assuming identical nonideal integrators, the relationship between $Q_{desired}$, the actual Q realized (Q_{actual}), and the integrator Q (Q_{int}) is, from (3.16)

$$\frac{1}{Q_{actual}} = \frac{1}{Q_{desired}} + \frac{2}{Q_{int}} \qquad (3.35)$$

The actual quality factor of the biquad is different from the desired one due to the excess phase shifts of the two integrators. This suggests the following strategy - if the value of Q_{int} is known *a priori*, then the biquad design can be *predistorted* for the particular value of integrator Q. That is, $Q_{desired}$ is modified to new value $Q_{predistort}$, according to

$$\frac{1}{Q_{predistort}} = \frac{1}{Q_{desired}} - \frac{2}{Q_{int}} \qquad (3.36)$$

This predistorting process ensures that the actual Q realized when finite quality integrators are used is equal to $Q_{desired}$.

If the integrator excess phase is due to a single high frequency pole at ω_1 (Figure 3.3), then,

$$Q_{int,lowend} \approx -\frac{\omega_1}{\omega_{lowend}} \approx 1/\Delta\phi_{lowend} \qquad (3.37)$$

where $\Delta\phi_{lowend}$ is the excess phase caused by the high frequency pole at ω_{lowend}. Consider now what happens when the center frequency is programmed to $\omega_{highend}$. As a numerical example, let $\omega_{highend} = 3\omega_{lowend}$. If the high frequency pole has not changed in the programming process (as is the case in most filter designs),

$$Q_{int,highend} \approx -\frac{\omega_1}{\omega_{highend}} \approx \frac{1}{3} Q_{int,lowend} \qquad (3.38)$$

If the filter design was predistorted at the low end to yield the correct response due to finite $Q_{int,lowend}$,

$$\frac{1}{Q_{predistort}} = \frac{1}{Q_{desired}} - \frac{2}{Q_{int,lowend}} \qquad (3.39)$$

At the high end frequency setting, we will have

$$\frac{1}{Q_{actual}} = \frac{1}{Q_{desired}} - \frac{2}{Q_{int,lowend}} + \frac{2}{Q_{int,highend}} \qquad (3.40)$$

Thus, the quality factor of the biquad at the low and high end center frequency settings is different, causing a change in the relative shape of the filter frequency response. This effect is especially severe for higher order filters, and/or ones which are supposed to have accurate group delay characteristics. From (3.40), we see two possible solutions to this problem.

- Make $Q_{int,lowend}$ and $Q_{int,highend}$ very large - this is the prevalent technique, and is attempted by trying to push high frequency poles as far away from $\omega_{highend}$ as possible. This approach generally has serious power and noise penalties, especially in high frequency filters.

- Make $Q_{int,lowend}$ and $Q_{int,highend}$ **equal** - this ensures that $Q_{actual} = Q_{desired}$ at the low end *and* high end settings. This is the approach taken in this book.

In more general terms, if the relative shape of the filter response has to be maintained over the programming range, the quality factor of the integrator must remain the same across the entire range of center frequencies desired. In terms of phase error, the programmable integrator must have *the same excess phase regardless of the set unity gain frequency*. In our example, ω_1 must be pushed to $3\omega_1$ at the high end setting. This would ensure that $Q_{int,lowend} = Q_{int,highend}$.

It is useful and informative to see the effect of a variable integrator phase error in a programmable filter. For illustrative purposes, we choose a fifth order equiripple group delay approximation. Figure 3.22 shows the normalized group delay response at two different bandwidth settings for the filter. The integrators used in the design of this filter are such that they have a single high frequency pole, which remains fixed with bandwidth programming. We see that at the low end bandwidth setting, the response is as desired (thanks to predistortion), but at the high end, there are serious aberrations in the group delay (due to the change in Q_{int}). The relative "peaking" of the group delay at higher bandwidths is conventionally tackled by moving parasitic poles farther away from the filter band edge (thereby reducing the *variation in* $\Delta \phi$ *over the programming range*).

Figure 3.23 shows the response of the filter which uses integrators that have a constant phase error. The shape of the response remains the same at the low and high ends of the programming range.

A Review of Integrator Architectures   59

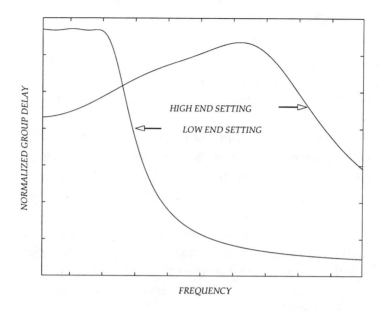

Figure 3.22. Aberrations in filter response at high bandwidth settings due to variations in phase error.

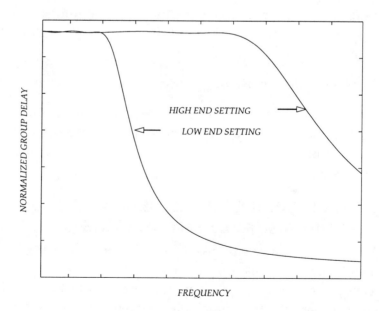

Figure 3.23. Response of a programmable filter, where the integrators have a constant phase error, independent of their unity gain frequencies.

7.2 MAINTAINING DYNAMIC RANGE IN WIDELY PROGRAMMABLE FILTERS

In this work, we define the dynamic range of the filter as the ratio of the maximal signal power that can be handled by the filter (for a given value of distortion) to the filter output noise power. It can be shown that the dynamic range of a filter is proportional to the dynamic range of the integrators used in the design [34]. When a programmable filter is designed, it is necessary to *maintain* the same dynamic range as the bandwidth is changed. Moreover, in many practical applications, it is desirable that the maximum signal handling capability of the filter does not change as the filter is programmed. To make this possible, the integrators used the filter design must have the same signal handling capability and the same output noise for all programmed unity gain frequencies. Ways to achieve this are discussed in the rest of this book.

7.3 SUMMARY

From the discussion in the preceding subsections, we conclude that the desirable properties of a good programmable integrator are

- Constant phase error across the tuning range - this is necessary to ensure that the relative shape of the filter response stays the same regardless of the set bandwidth.

- Constant output noise power across the tuning range.

- Constant maximum signal handling capacity across the tuning range - alongwith the output noise property above, this ensures that the filter dynamic range does not degrade as its bandwidth is changed.

8. APPROACHES TO THE DESIGN OF WIDELY PROGRAMMABLE INTEGRATORS

A brief discussion of the choices in programmable integrator design is given in this section. We only examine the specific case of Gm-C filters due to their importance in practical CMOS high frequency design. Any filter pole frequency is of the form G_m/C. This means that there are two fundamental ways of programming the poles of the filter - one could keep G_m constant and vary C,

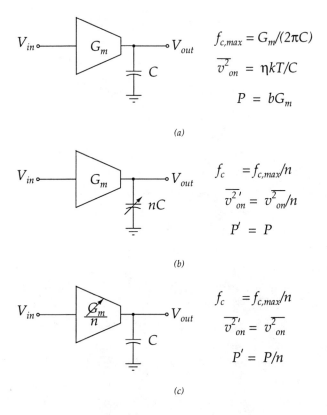

Figure 3.24. (a) A Gm-C integrator designed to achieve the maximum cutoff frequency in a programmable range; (b) Lowering the cutoff frequency of (a) by increasing the capacitance ("constant G_m" approach); (c) Lowering the cutoff frequency of (a) by decreasing the transconductance ("constant C" approach.)

or viceversa. These are termed "constant-G_m" and "constant-C" respectively. Since each of these approaches largely effect filter noise and power dissipation, these issues are examined first.

We consider a $G_m - C$ integrator which must be programmed over a wide range of unity-gain frequencies, the largest of which is $f_{c,max}$. We assume that an optimized design for the toughest case (the largest cutoff frequency) has been achieved, as in Figure 3.24(a). The relation of this frequency to component values, total integrated mean square output noise (excluding $1/f$ noise), and power dissipation is given next to the figure; it is assumed that the noise current power spectral density is $4kT\eta G_m$, where η is the excess noise factor of the transconductor [34]. Also, as is often the case in optimized transconductor design, the power dissipation is assumed to be proportional to G_m, with a constant of proportionality b, depending on design details. For a given noise spec $\overline{v_{on}^2}$, the value of C is set, which from the top relation in the figure sets G_m and thus P.

Figures 3.24(b) and (c) show the corresponding cutoff frequency, mean squared noise and power dissipation (denoted by primes) for the "constant-G_m" and "constant-C" approaches, in terms of those at $f_c = f_{c,max}$. From these results it follows that, as f_c is programmed, the noise, total capacitance area and power dissipation will vary as shown in Figure 3.25. It is clear that the constant-G_m design can result in unnecessarily large capacitance in order to achieve low cutoff frequencies. In low-noise applications, where chip area can be capacitance-limited, this can be a serious problem. The fact that very low noise is achieved by the constant-G_m design for low values of f_c is irrelevant, as far as the spec (indicated on the top plot) is concerned; the total capacitance cannot be decreased, as then the entire noise plot will rise, and then the spec will not be satisfied at high frequencies. In contrast, the constant-C design needs no capacitance increase, and the noise spec is still satisfied everywhere. Finally, since the low f_c values are achieved by decreasing G_m, a lower power dissipation can be achieved at low f_c values, depending on the transconductor design (in the technique proposed in Chapter 4, this last advantage is given up to achieve design robustness). Due to the above observations, the constant-C scaling technique is the preferred approach, when integrator output noise power is considered.

Signal handling capability and phase error are depend strongly on the particular transconductor topology in use. It turns out that in *conventional Gm – C techniques*, as we will illustrate with an example,

- At the low end bandwidth, integrator dynamic range deteriorates while it is easy to maintain an accurate frequency response.

- At the high end bandwidth, frequency response accuracy tends to deteriorate.

Consider the Gm-C integrator shown in Figure 3.26 [31] [61] [62] [9] [54]. We assume that the transconductances of all transistors (M1

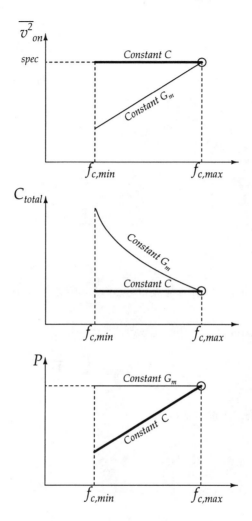

Figure 3.25. Variation of mean squared noise, total capacitance and power dissipation versus cutoff frequency for programmable integrators using the "constant G_m" approach (thin line) and the "constant C" approach (thick line). The circle indicates a design optimized for the maximum f_c setting.

through M6) are identical and equal to g_m. (In practice, this will approximately be true). The body terminals of all the devices are assumed to be connected to their sources for simplicity. C_i represents the integrating capacitance. C_p is a parasitic capacitance that is

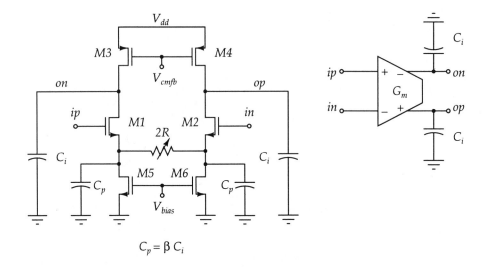

Figure 3.26. A conventional Gm-C integrator.

neglected for now. It can be shown that

$$G_m = \frac{1}{2R}\left(\frac{g_m R}{g_m R + 1}\right) \qquad (3.41)$$

Typically $g_m R \gg 1$ and hence

$$G_m \approx \frac{1}{2R} \qquad (3.42)$$

The unity gain frequency is given by

$$\omega_o \approx \left(\frac{1}{RC_i}\right) \qquad (3.43)$$

and is programmed by changing the value of the degenerating resistor $2R$ (typically implemented as a MOSFET operating in the triode region, as discussed earlier in this chapter). Let us assume that ω_o has to be programmed by a factor of 5 from ω_{min} to $5\omega_{min}$

Further, let us restrict the minimum value of $g_m R$ to 10. The input referred noise voltage spectral density for the transconductor, including the contribution of the current sources, can be shown to be

$$\frac{\overline{v_n^2}}{\Delta f} \approx 4kT\left(\frac{2}{3}\right)2R\,(2g_m R) \qquad (3.44)$$

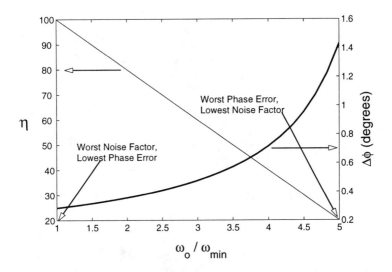

Figure 3.27. η and $\Delta\phi$ as a function of ω_o.

The excess noise factor of the transconductor is therefore, by comparison to (3.25),

$$\eta = \frac{4}{3} g_m R \qquad (3.45)$$

If C_p is not neglected, it can be shown that the integrator transfer function has a parasitic pole-zero pair. The effect of the zero can be cancelled using predistortion of the filter transfer function [43]. What remains then, is the excess phase caused by the parasitic pole at

$$\omega_p \approx \left(\frac{g_m}{C_p}\right) \qquad (3.46)$$

which causes a phase lag in the integrator transfer function at its unity gain frequency approximately given by

$$\Delta\phi \approx \tan^{-1}\left(\frac{C_p}{C_i g_m R}\right) \qquad (3.47)$$

Figure 3.27 shows η and $\Delta\phi$ as ω_o is programmed from ω_{min} to $5\omega_{min}$. The maximum signal handling capability is assumed constant irrespective of the value of $2R$. Since the total output noise power of a Gm-C integrator is directly proportional to the excess noise factor and

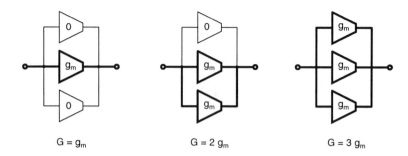

Figure 3.28. A programmable transconductor using multiple switchable unit transconductors.

inversely dependent on the value of the integrating capacitance, we conclude that at the low end bandwidth setting the dynamic range is worst because η is the highest. At high frequency settings, η is lower but there is a large phase error. Thus, this circuit is plagued by varying excess phase shift and varying excess noise factor, each of these achieving their worst-case values at opposite extremes of the programmable frequency range. This makes design difficult, and results in sub-optimum programmable filters.

8.1 VARIABLE G_M VERSUS MULTIPLE UNIT CELLS

An alternative design technique, which does not suffer from the problems discussed above, is to compose G_m out of optimized unit transconductor cells, and switch these in parallel to achieve the desired programmable transconductance values. An example of a transconductor tunable by a factor of three is shown in Figure 3.28. If continuous tuning is needed in between the discrete steps, one can control the G_m values over a small range using well known continuous tuning techniques. Since this range will be very small, the variability of performance over this range will be small too. This case is not considered below.

8.2 EFFECT OF PARASITIC CAPACITANCES ON PROGRAMMABILITY

When switchable transconductor cells are used, a significant problem arises [35]. To illustrate this problem, we consider the case of a second-order filter, as shown in Figure 3.29(a). The pole frequency and quality factor of this filter are as follows :

$$f_o = \frac{1}{2\pi}\sqrt{\frac{G_{m1}G_{m2}}{C_a C_b}} \tag{3.48}$$

$$Q = \sqrt{\frac{G_{m1}}{G_{m2}}\frac{C_a}{C_b}} \tag{3.49}$$

To make the pole frequency programmable, G_{m1} and G_{m2} are commonly varied by the same factor, and thus $\frac{G_{m1}}{G_{m2}}$ remains fixed. However, if switchable unit cells are used to implement the programmable G_{m1} and G_{m2}, then each time such cells are switched in and out, the total value of the parasitic capacitance at each node changes, in a manner that is very difficult to control. This is illustrated in Figure 3.29(b), where the arrows indicated elements that are variable (intentionally or unintentionally). C_{i1} and C_{o1} are the input and output parasitic capacitances of transconductors G_{m1}, and C_{i2} and C_{o2} are the corresponding quantities for transconductors G_{m2}. The pole frequency and quality factor of this filter now become

$$f_o = \frac{1}{2\pi}\sqrt{\frac{G_{m1}G_{m2}}{(C_a + 2C_{o1} + 2C_{i2} + C_{o2})(C_b + C_{i1} + C_{o2})}} \tag{3.50}$$

$$Q = \sqrt{\frac{G_{m1}}{G_{m2}}\frac{(C_a + C_{o1} + 2C_{i2} + C_{o2})}{(C_b + C_{i1} + C_{o2})}} \tag{3.51}$$

The variation of parasitics causes two problems: the pole frequency will not change in proportion to the stepped G_{m1} and G_{m2}; and the changing parasitics will change Q, resulting in response shape errors. The latter effect is especially severe for higher order filters, and/or ones which are supposed to have accurate group delay characteristics. It is possible that a design which meets the specs at the highest pole frequency will not meet them at lower pole frequencies, due to the variability of parasitics as transconductances are switched in and out of the filter. It is thus seen that it is not enough to achieve a working design at the highest pole frequency of interest, and then hope that there will not be any problems at lower pole frequency settings.

Consider now the situation in Figure 3.29(c). Assume that all parasitic capacitances can be kept fixed, and only the G_ms are variable.

68 HIGH FREQUENCY CONTINUOUS TIME FILTERS

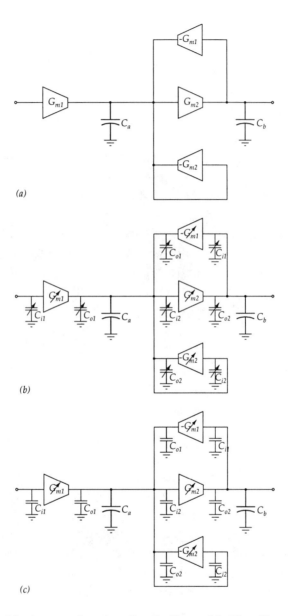

Figure 3.29. (a) A second order Gm-C filter; (b) The filter in (a), made programmable by varying transconductances; the parasitic input and output capacitances also vary, as indicated by arrows; (c) The filter in (a), assuming all parasitics remain fixed while transconductances are varied.

Once the fixed value of parasitics has been incorporated into the design, it is seen from (3.48) that the pole frequency will now change in proportion to the G_m's, and Q will be fixed as seen from (3.51). In this figure, notice that *every* transconductance in the network is multiplied by a programming factor, while keeping *every* capacitance the same (*including* the parasitic ones). The next two chapters cover techniques that make possible the reliable implementation of this principle.

The preceding discussion has focussed on a biquad based filter implementation, but the principles are general and can be extended to arbitrary networks of transconductances, conductances and capacitors. In fact, interesting large signal properties of programmable G_m-C networks built with nonlinear elements can be derived. A more rigorous analysis of noise, signal handling capability and dynamic range in "constant-G_m" and "constant-C" networks is the subject of the next chapter.

9. CONCLUSIONS

After a study of various possible architectures we conclude that Gm-grounded-C filters show the most promise for realizing very high frequency continuous time filters in digital CMOS processes. A comparison of the many different transconductor circuits suggests that the differential pair is probably best suited for high frequency work, as long as the linearity requirements are modest. We have also identified the key problems associated with conventional bandwidth programming techniques - dynamic range is the most difficult to maintain at low end frequency settings and frequency response accuracy degrades at high frequencies. An approach to mitigate these problems, that uses optimized multiple switchable unit transconductor cells in parallel, was suggested. This approach will be discussed in detail in the rest of this book.

Chapter 4

TIME SCALING IN ELECTRICAL NETWORKS

1. INTRODUCTION

The constant-G_m and constant-C techniques discussed in the last chapter are well known cases of frequency scaled electrical networks. Scaling of the frequency and time axes are related through the Fourier transform, but we prefer to use the term "time-scaling" rather than "frequency-scaling" because the former term is far more general and can be applied to nonlinear circuits. In this chapter, we discuss the concept of time scaling and its application in the design of programmable analog filters. We first review the correspondence between scaling of the time and frequency axes for the case of linear networks. The constant-G_m and constant-C scaling techniques (from now on referred to as constant conductance and constant capacitance scaling respectively) are discussed along with their noise properties. We then extend the concept of time-scaling to nonlinear networks and derive some large signal properties of scaled nonlinear networks [36]. From these results, and the noise behavior of scaled networks, we show that constant-capacitance scaled filters are an optimal choice for designing widely programmable analog filters. Finally, CMOS implementations of scaled filters are presented. Throughout our discussion, $1/f$ noise is neglected; only white noise sources are considered.

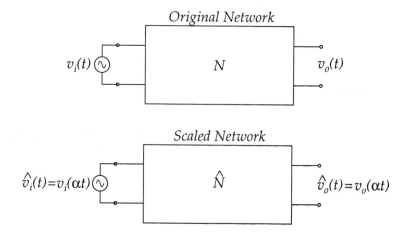

Figure 4.1. Definition of time scaling.

2. TIME SCALING: DEFINITION

Consider an initially relaxed network $\mathcal{N}$. For the time being we consider a single input, single output network. Let an arbitrary input voltage $v_i(t)$ produce an output voltage $v_o(t)$. Further, let us assume the existence of a network $\hat{\mathcal{N}}$ satisfying the following condition: **An input $\hat{v}_i(t) = v_i(\alpha t)$ results in an output $\hat{v}_o(t) = v_o(\alpha t)$.** The only restriction on α is that it be positive. Then, we call $\hat{\mathcal{N}}$ the time-scaled version of $\mathcal{N}$, with a scaling factor α. This is denoted in Figure 4.1. We introduce the notation shown below to denote the relationship between a network and its time-scaled cousin.

$$\mathcal{N} \xrightarrow{t|\alpha t} \hat{\mathcal{N}} \qquad (4.1)$$

3. THE LINEAR CASE

If $\mathcal{N}$ is a linear network, then a correspondence can be established between scaling of the time and frequency axes. In Figure 4.1, let the Fourier transforms of $v_i(t)$ and $v_o(t)$ be $V_i(f)$ and $V_o(f)$ respectively. The transfer function of $\mathcal{N}$ is

$$H(f) = \frac{V_o(f)}{V_i(f)} \qquad (4.2)$$

The transfer function of $\hat{\mathcal{N}}$ can be calculated from

$$\hat{H}(f) = \frac{\mathcal{F}(\hat{v}_o(t))}{\mathcal{F}(\hat{v}_i(t))} = \frac{\mathcal{F}(v_o(\alpha t))}{\mathcal{F}(v_i(\alpha t))} \quad (4.3)$$

Since

$$\mathcal{F}[v_i(\alpha t)] = \frac{1}{\alpha} V_i(f/\alpha) \quad (4.4)$$

$$\mathcal{F}[v_o(\alpha t)] = \frac{1}{\alpha} V_o(f/\alpha) \quad (4.5)$$

we get

$$\hat{H}(f) = \frac{V_o(f/\alpha)}{V_i(f/\alpha)} \quad (4.6)$$

Comparing (4.6) with (4.2), we can write

$$\boxed{\hat{H}(f) = H(f/\alpha)} \quad (4.7)$$

From this we see that stretching of the time axis is equivalent to "compressing" the frequency axis and vice-versa. **The relative shape of the magnitude response remains the same.** Table 4.1 reflects the basic relationships obeyed by scaled networks.

Parameter	$\mathcal{N}$	$\hat{\mathcal{N}}$
Input voltage	$v_i(t)$	$v_i(\alpha t)$
Output voltage	$v_o(t)$	$v_o(\alpha t)$
Frequency Response	$H(f)$	$H(f/\alpha)$
Magnitude Response	$|H(f)|$	$|H(f/\alpha)|$
Phase Response	$\angle H(f)$	$\angle H(f/\alpha)$
Group Delay	$\tau(f)$	$\frac{1}{\alpha}\tau(f/\alpha)$

Table 4.1. Basic properties of scaled networks, $\mathcal{N} \xrightarrow{t/\alpha t} \hat{\mathcal{N}}$.

In (trans)conductance-capacitance networks, two methods of realizing a scaled network become obvious. We choose to illustrate with a simple example. Consider the network shown in Figure 4.2. The

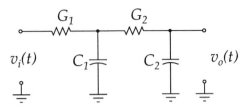

Figure 4.2. An example network.

transfer function is

$$H(jf) = 1/\left[1 + jf\left(\frac{C_1}{G_1} + \frac{C_2}{G_2} + \frac{C_2}{G_1}\right) + (jf)^2 \frac{C_1 C_2}{G_1 G_2}\right] \quad (4.8)$$

Notice in (4.8) that the frequency variable f is always multiplied by a factor of the form $\sum \frac{C}{G}$. For instance, the middle term in the denominator is

$$jf\left(\frac{C_1}{G_1} + \frac{C_2}{G_2} + \frac{C_2}{G_1}\right) = jf \times \left(\frac{C_1}{G_1} + \frac{C_2}{G_2} + \frac{C_2}{G_1}\right) \quad (4.9)$$

while the second term is

$$(jf)^2 \frac{C_1 C_2}{G_1 G_2} = \left(jf \times \frac{C_1}{G_1}\right)\left(jf \times \frac{C_2}{G_2}\right) \quad (4.10)$$

This suggests two straightforward ways of accomplishing time scaling.

- Multiply all capacitances by $1/\alpha$, while keeping all conductances and transconductances unchanged. We call this **constant-conductance scaling**.

- Multiply all conductances and transconductances by α, while keeping all capacitances unchanged. We call this **constant-capacitance scaling**.

In Figure 4.3, we illustrate both scaling techniques for the network of Figure 4.2. It is obvious that the above discussion can be extended to arbitrary networks of (trans)conductances and capacitors (Figure 4.4). For a rigorous proof, the reader is referred to [37]. Both scaling strategies satisfy the properties listed in Table 4.1. The above two scaling techniques are not unique. For instance, one could easily think of a strategy where all conductances are scaled by $\sqrt{\alpha}$ while all capacitors are scaled by $\sqrt{1/\alpha}$.

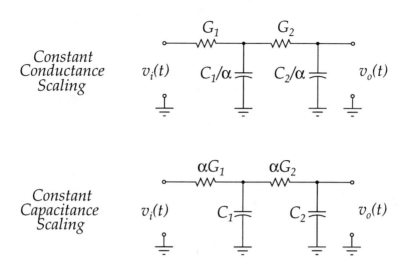

Figure 4.3. Illustration of constant-conductance and constant-capacitance scaling for the network of Figure 4.2.

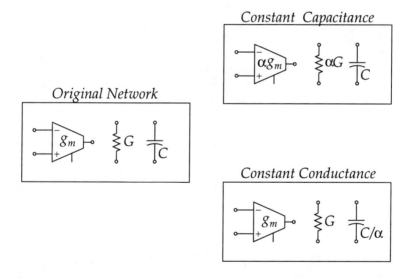

Figure 4.4. Illustration of constant-conductance and constant-capacitance scaling for a general (trans)conductance-capacitance network.

4. NOISE PROPERTIES OF SCALED NETWORKS

In this section, we analyze the noise properties of constant conductance and constant-capacitance scaled networks [38]. Let us assume that the noise of every transconductance and conductance can

76 HIGH FREQUENCY CONTINUOUS TIME FILTERS

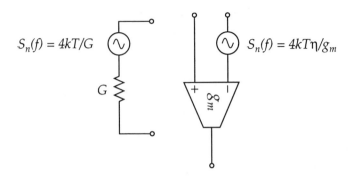

Figure 4.5. Noise model for the conductances and transconductors.

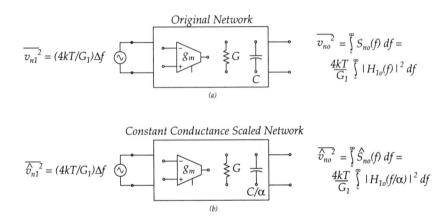

Figure 4.6. Noise properties of constant-conductance scaled networks.

be represented by the models shown in Figure 4.5. k is Boltzmann's constant and T is the absolute temperature. η is the excess noise factor of the transconductor. $1/f$ noise is neglected.

4.1 NOISE IN CONSTANT-CONDUCTANCE NETWORKS

Consider Figure 4.6(a), where we isolate a single noise source v_{n1} from the network. We are interested in finding the output noise spectral density $S_{no}(f)$ and the integrated output noise voltage $\overline{v_{no}^2}$. We denote the transfer function from v_{n1} to v_o as $H_{1o}(f)$. We can write

$$S_{no}(f) = |H_{1o}(f)|^2 \times \frac{4kT}{G_1} \qquad (4.11)$$

$$\overline{v_{no}^2} = \int_0^\infty S_{no}(f)df = \int_0^\infty |H_{10}(f)|^2 \times \frac{4kT}{G_1}df = \frac{4kT}{G_1}\int_0^\infty |H_{10}(f)|^2 df \quad (4.12)$$

For the scaled network (Figure 4.6(b)), the relations corresponding to (4.11) and (4.12) are

$$\hat{S}_{no}(f) = |H_{10}(f/\alpha)|^2 \times \frac{4kT}{G_1} \quad (4.13)$$

$$\overline{\hat{v}_{no}^2} = \int_0^\infty \hat{S}_{no}(f)df = \int_0^\infty |H_{10}(f/\alpha)|^2 \times \frac{4kT}{G_1}df = \alpha\frac{4kT}{G_1}\int_0^\infty |H_{10}(f/\alpha)|^2 d(f/\alpha) \quad (4.14)$$

From (4.11) and (4.13), we get

$$\hat{S}_{no}(f) = S_{no}(f/\alpha) \quad (4.15)$$

Comparing (4.14) and (4.12), we see that

$$\overline{v_{no}^2} = \alpha\overline{\hat{v}_{no}^2} \quad (4.16)$$

Following the same reasoning as above and utilizing superposition, it is easy to extend the results to multiple noise sources and conclude that the integrated output noise of a constant-conductance scaled network is directly proportional to the scaling factor α. The noise properties are summarized in Table 4.2.

Parameter	$\mathcal{N}$	$\hat{\mathcal{N}}$
Noise Spectral Density	$S_{no}(f)$	$S_{no}(f/\alpha)$
Integrated Output Noise	$\overline{v_{no}^2}$	$\alpha\overline{v_{no}^2}$
Frequency Response	$H(f)$	$H(f/\alpha)$

Table 4.2. Noise properties of constant-conductance scaled networks, $\mathcal{N} \xrightarrow{t|\alpha t} \hat{\mathcal{N}}$.

Thus, a filter designed using constant-conductance-scaling principles will have a smaller output noise for small values of α, corresponding to a higher value of total circuit capacitance used.

4.2 NOISE IN CONSTANT-CAPACITANCE NETWORKS

Consider Figure 4.7(a). To find the output noise spectral density and integrated output noise, we use (4.11) and (4.12). For the scaled

78 HIGH FREQUENCY CONTINUOUS TIME FILTERS

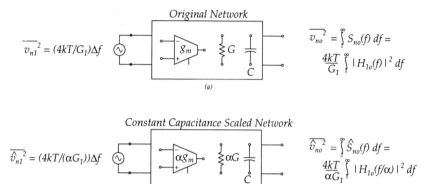

Figure 4.7. Noise properties of constant-capacitance scaled networks.

network (Figure 4.7(b)), the relations corresponding to (4.11) and (4.12) are

$$\hat{S}_{no}(f) = |H_{10}(f/\alpha)|^2 \times \frac{4kT}{\alpha G_1} \quad (4.17)$$

$$\overline{\hat{v}_{no}^2} = \int_0^\infty \hat{S}_{no}(f) df = \int_0^\infty |H_{10}(f/\alpha)|^2 \times \frac{4kT}{\alpha G_1} df = \frac{4kT}{G_1} \int_0^\infty |H_{10}(f/\alpha)|^2 d(f/\alpha) \quad (4.18)$$

From (4.11) and (4.17), we get

$$\hat{S}_{no}(f) = \frac{1}{\alpha} S_{no}(f/\alpha) \quad (4.19)$$

Comparing (4.18) and (4.12), we see that

$$\overline{v_{no}^2} = \overline{\hat{v}_{no}^2} \quad (4.20)$$

Following the same reasoning as above and utilizing superposition, it is easy to extend the results to multiple noise sources and conclude that **the integrated output noise of a constant-capacitance scaled network is independent of the scaling factor** α. (excluding $1/f$ noise). The noise properties are summarized in Table 4.3.

A filter designed using constant-capacitance-scaling principles will have the same RMS output noise irrespective of its bandwidth, whereas in the case of constant conductance scaling, the total noise power varies with the scaling factor α. We now consider a programmable filter, with α settable between the values of 1 and α_{max}, where $\alpha_{max} \gg 1$. If the design of this filter is based on constant conductance scaling, it must be such that the filter meets the noise

Parameter	$\mathcal{N}$	$\hat{\mathcal{N}}$
Noise Spectral Density	$S_{no}(f)$	$\frac{1}{\alpha}S_{no}(f/\alpha)$
Integrated Output Noise	$\overline{v_{no}^2}$	$\overline{v_{no}^2}$
Frequency Response	$H(f)$	$H(f/\alpha)$

Table 4.3. Noise properties of constant-capacitance scaled networks, $\mathcal{N} \xrightarrow{t|\alpha t} \hat{\mathcal{N}}$.

specification at the setting $\alpha = \alpha_{max}$ (see Table 4.2). As a result, the filter will be grossly over-designed at lower settings of α.

In contrast to this, if the programmable filter is designed using constant capacitance scaling, the output noise will be independent of the scaling factor (Table 4.3). Thus, if the filter is designed optimally for one setting of α, it will remain optimal for all other settings, and no over-design will be needed.

It has been shown elsewhere [34] that the output noise power is inversely proportional to the total filter capacitance. Combining this fact with the above observations, it is easy to see that a constant conductance programmable filter will need α_{max} times more capacitance than a constant capacitance design. This is why programmable filters should be designed using constant capacitance scaling. It can also be shown that, although the two filters will have the same total power dissipation at the most wideband setting, in the constant capacitance case power dissipation will decrease as α is decreased, whereas in the constant conductance case it will remain constant, independent of α.

5. EXTENSION OF SCALING TO THE NONLINEAR CASE

The concepts of frequency response and transfer function do not exist, strictly speaking, for nonlinear systems, hence we can only talk about time scaling. Consider the case when the (trans)conductors and capacitors are nonlinear. The input-output relation for a nonlinear transconductor is denoted as

$$i = f_{abc}(v_a, v_b) \tag{4.21}$$

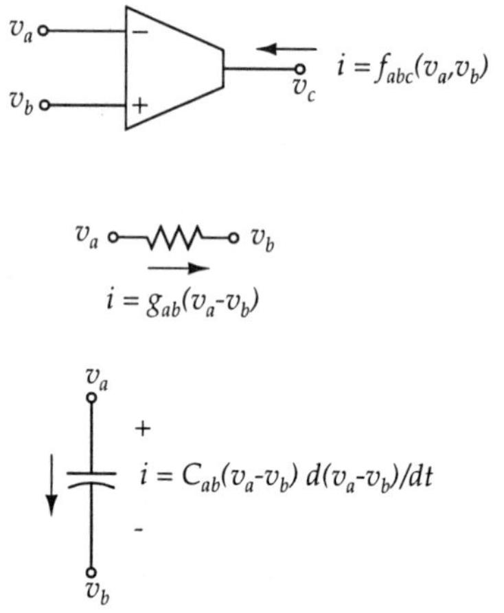

Figure 4.8. Model for nonlinear (trans)conductors and capacitors.

where v_a, v_b and v_c represent the voltages at the input and output nodes as shown in Figure 4.8. Similar $i - v$ relations can be written for the conductors and the nonlinear capacitors. Consider now a general network consisting of these elements. We resort to the Modified Nodal Analysis (MNA) formulation of writing the network equations [39]. The MNA equations are written using the following steps (the network is assumed to have n nodes).

- Choose a datum node.
- For $k = 1, 2, \ldots, n - 1$, write Kirchhoff's Current Law (KCL) for node k using the node-to-datum voltages as variables, keeping in mind that if one or more branches connected to the node k are not voltage-controlled, then the corresponding branch current is entered in the node equation and the corresponding branch equation is appended to the $n - 1$ node equations.

While we consider only a single-input-single-output network here, the results we derive can easily be generalized to multiple input, multiple output circuits. Following the MNA formulation described above, we would have, for the network of Figure 4.9(a)

$$\sum_p C_{1p}(v_1(t) - v_p(t)) \frac{d(v_1(t) - v_p(t))}{dt} + \sum_{q,r} f_{qr1}(v_q(t), v_r(t)) + \sum_u g_{1u}(v_1(t) - v_u(t)) = 0$$

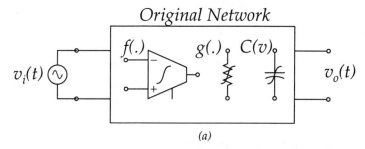

(a)

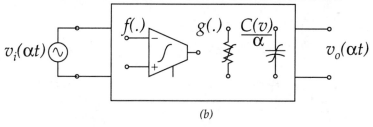

(b)

Figure 4.9. Constant-conductance scaling in nonlinear networks.

$$\vdots$$

$$\sum_p C_{kp}(v_k(t) - v_p(t)) \frac{d(v_k(t) - v_p(t))}{dt} +$$

$$\sum_{q,r} f_{qrk}(v_q(t), v_r(t)) + \sum_u g_{ku}(v_k(t) - v_u(t)) + i_{v_i}(t) = 0$$

$$\vdots$$

$$\sum_p C_{(n-1)p}(v_{n-1}(t) - v_p(t)) \frac{d(v_{n-1}(t) - v_p(t))}{dt} +$$

$$\sum_{q,r} f_{qr(n-1)}(v_q(t), v_r(t)) + \sum_u g_{(n-1)u}(v_{n-1}(t) - v_u(t)) = 0$$

$$v_k(t) = v_i(t)$$

(4.22)

where $v_i(t)$ and $i_{v_i}(t)$ represent the input voltage source and the current through it respectively. One terminal of the input voltage source has been chosen as the datum node and the other terminal is chosen to be the kth node. Notice that there are n equations in all - the n unknowns are the $(n-1)$ node voltages and $i_{v_i}(t)$. If $i_{v_i}(t)$ is not required, the kth equation can be omitted. Using the resulting

82 HIGH FREQUENCY CONTINUOUS TIME FILTERS

$n-1$ equations, alongwith the capacitor voltages at $t=0$ as initial conditions, the $n-1$ node voltages can be solved for. Once this is done, it is trivial to compute $i_{v_i}(t)$ from the kth equation.

Let us replace t in the set of equations (4.22) by αt. We obtain

$$\sum_p C_{1p}(v_1(\alpha t) - v_p(\alpha t)) \frac{d(v_1(\alpha t) - v_p(\alpha t))}{d(\alpha t)} +$$
$$\sum_{q,r} f_{qr1}(v_q(\alpha t), v_r(\alpha t)) + \sum_u g_{1u}(v_1(\alpha t) - v_u(\alpha t)) = 0$$

$$\vdots$$

$$\sum_p C_{kp}(v_k(\alpha t) - v_p(\alpha t)) \frac{d(v_k(\alpha t) - v_p(\alpha t))}{d(\alpha t)} +$$
$$\sum_{q,r} f_{qrk}(v_q(\alpha t), v_r(\alpha t)) + \sum_u g_{ku}(v_k(\alpha t) - v_u(\alpha t)) + i_{v_i}(\alpha t) = 0$$

$$\vdots$$

$$\sum_p C_{(n-1)p}(v_{n-1}(\alpha t) - v_p(\alpha t)) \frac{d(v_{n-1}(\alpha t) - v_p(\alpha t))}{d(\alpha t)} +$$
$$\sum_{q,r} f_{qr(n-1)}(v_q(\alpha t), v_r(\alpha t)) + \sum_u g_{(n-1)u}(v_{n-1}(\alpha t) - v_u(\alpha t)) = 0$$

$$v_k(\alpha t) = v_i(\alpha t)$$
(4.23)

Equations (4.23) can be rewritten as

$$\sum_p \frac{C_{1p}(v_1(\alpha t) - v_p(\alpha t))}{\alpha} \frac{d(v_1(\alpha t) - v_p(\alpha t))}{dt} +$$
$$\sum_{q,r} f_{qr1}(v_q(\alpha t), v_r(\alpha t)) + \sum_u g_{1u}(v_1(\alpha t) - v_u(\alpha t)) = 0$$

$$\vdots$$

$$\sum_p \frac{C_{kp}(v_k(\alpha t) - v_p(\alpha t))}{\alpha} \frac{d(v_k(\alpha t) - v_p(\alpha t))}{dt} +$$
$$\sum_{q,r} f_{qrk}(v_q(\alpha t), v_r(\alpha t)) + \sum_u g_{ku}(v_k(\alpha t) - v_u(\alpha t)) + i_{v_i}(\alpha t) = 0$$

$$\vdots$$

$$\sum_p \frac{C_{(n-1)p}(v_{n-1}(\alpha t) - v_p(\alpha t))}{\alpha} \frac{d(v_{n-1}(\alpha t) - v_p(\alpha t))}{dt} +$$

$$\sum_{q,r} f_{qr(n-1)}(v_q(\alpha t), v_r(\alpha t)) + \sum_u g_{(n-1)u}(v_{n-1}(\alpha t) - v_u(\alpha t)) = 0$$

$$v_k(\alpha t) = v_i(\alpha t) \quad (4.24)$$

Notice that these are the set of equations one would get for the network of Figure 4.9(b) - where all the (trans)conductances are the same as in the original network, while all the capacitors have been scaled by the factor $\frac{1}{\alpha}$. We refer to this as the "constant-conductance" scaled network. Let $v_m(t)$ be a node voltage in the original network of Figure 4.9(a) and $\hat{v}_m(t)$ be the corresponding node voltage in the scaled network of Figure 4.9(b). The set $\{v_m(t)\}$ is the solution of (4.22), when the original network is excited by an input $v_i(t)$. Thus, from (4.24), the set of node voltages for the scaled network will be $\{\hat{v}_m(t) = v_m(\alpha t)\}$ if that network is excited by the input $\hat{v}_i(t) = v_i(\alpha t)$. As in the linear case, the relationship between the outputs of the original and scaled network reads : **If an input $v_i(t)$ to the original network produces an output $v_o(t)$, an input $v_i(\alpha t)$ applied to the scaled network produces an output $v_o(\alpha t)$.** Note that the scaled network must have the same set of initial conditions as the unscaled network for the above to hold.

Alternatively, (4.23) can be rewritten as

$$\sum_p C_{1p}(v_1(\alpha t) - v_p(\alpha t)) \frac{d(v_1(\alpha t) - v_p(\alpha t))}{dt} +$$

$$\sum_{q,r} \alpha f_{qr1}(v_q(\alpha t), v_r(\alpha t)) + \sum_u \alpha g_{1u}(v_1(\alpha t) - v_u(\alpha t)) = 0$$

$$\vdots$$

$$\sum_p C_{kp}(v_k(\alpha t) - v_p(\alpha t)) \frac{d(v_k(\alpha t) - v_p(\alpha t))}{dt} +$$

$$\sum_{q,r} \alpha f_{qrk}(v_q(\alpha t), v_r(\alpha t)) + \sum_u \alpha g_{ku}(v_k(\alpha t) - v_u(\alpha t)) + \alpha i_{v_i}(\alpha t) = 0$$

$$\vdots$$

$$\sum_p C_{(n-1)p}(v_{n-1}(\alpha t) - v_p(\alpha t)) \frac{d(v_{n-1}(\alpha t) - v_p(\alpha t))}{dt} +$$

$$\sum_{q,r} \alpha f_{qr(n-1)}(v_q(\alpha t), v_r(\alpha t)) + \sum_u \alpha g_{(n-1)u}(v_{n-1}(\alpha t) - v_u(\alpha t)) = 0$$

$$v_k(\alpha t) = v_i(\alpha t)$$

84 HIGH FREQUENCY CONTINUOUS TIME FILTERS

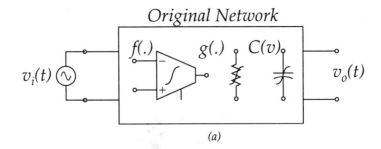

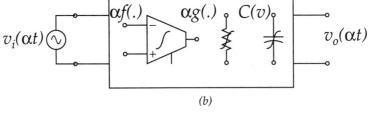

Figure 4.10. Constant-capacitance scaling in nonlinear networks.

$$(4.25)$$

Reasoning as above, we see that these equations are those that one would obtain for the network of Figure 4.10(b) - where all the capacitors are the same as in the original network, while all conductors and transconductors have been scaled by the factor α. We refer to this as the "constant-capacitance" scaled network.

As an illustration, consider the simple nonlinear network shown in Figure 4.11 (this is commonly known as a peak-detector). The networks are assumed to be relaxed initially. The diodes are modeled by the $i - v$ relationship (ϕ_t stands for the thermal voltage $\frac{kT}{q}$)

$$i = I_o \left[\exp\left(\frac{v}{\phi_t}\right) - 1 \right] \qquad (4.26)$$

Writing KCL for the network of Figure 4.11(a), we get

$$\frac{dv_{oa}(t)}{dt} = I_o \left[\exp\left(\frac{V_a \sin(t) - v_{oa}(t)}{\phi_t}\right) - 1 \right] - v_{oa}(t) \qquad (4.27)$$

Time Scaling in Electrical Networks   85

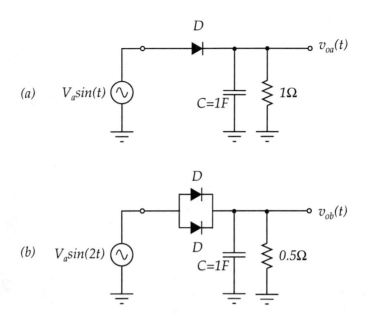

Figure 4.11. A network example to illustrate scaling - (a) Original network, (b) network constant-C scaled by $\alpha = 2$.

For the network of Figure 4.11(b), we get

$$\frac{dv_{ob}(t)}{dt} = 2I_o \left[\exp\left(\frac{V_a \sin(2t) - v_{ob}(t)}{\phi_t}\right) - 1 \right] - 2v_{ob}(t) \quad (4.28)$$

To see the relation between v_{ob} and v_{oa}, we proceed as follows. First, we replace t in (4.27) by $2t$. This gives

$$\frac{dv_{oa}(2t)}{d(2t)} = I_o \left[\exp\left(\frac{V_a \sin(2t) - v_{oa}(2t)}{\phi_t}\right) - 1 \right] - v_{oa}(2t) \quad (4.29)$$

This can be rewritten as

$$\frac{dv_{oa}(2t)}{dt} = 2I_o \left[\exp\left(\frac{V_a \sin(2t) - v_{oa}(2t)}{\phi_t}\right) - 1 \right] - 2v_{oa}(2t) \quad (4.30)$$

Comparing this with (4.28), and since $v_{ob}(0) = v_{oa}(0)$, we conclude that

$$v_{ob}(t) = v_{oa}(2t) \quad (4.31)$$

In Figure 4.12, we show the simulated results of the circuits of Figure 4.11. The top and middle plots show $v_{oa}(t)$ and $v_{ob}(t)$ respectively. In the bottom plot, $v_{oa}(t)$ and $v_{ob}(t/2)$ are overlaid. The two outputs lie on top of each other. This should come as no surprise.

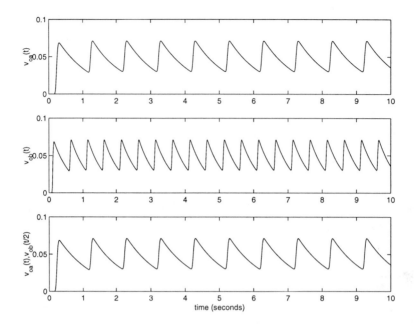

Figure 4.12. Top: Output waveform $v_{oa}(t)$, Middle: Output waveform $v_{ob}(t)$, Bottom: $v_{oa}(t)$ and $v_{ob}(t/2)$ overlaid on the same plot.

6. DISTORTION IN WEAKLY NONLINEAR SCALED FILTERS

Figure 4.13 shows two weakly nonlinear filters, where one is the constant-C scaled version of the other. The network of (a) is excited by a sinewave of amplitude A, and a frequency ω_o. We express the output as a Fourier Series

$$v_o(t) = \sum_n A_n \sin(n\omega_o t) + B_n \cos(n\omega_o t) \qquad (4.32)$$

Since the filter of Figure 4.13(b) is a scaled version with a scaling factor α, its output for an excitation $A \sin(\alpha \omega_o t)$ is

$$\hat{v}_o(t) = v_o(\alpha t) = \sum_n A_n \sin(n\alpha\omega_o t) + B_n \cos(n\alpha\omega_o t) \qquad (4.33)$$

Observe that the Fourier coefficients of the output remain the same for both the networks. This means that if the original filter has a Total Harmonic Distortion (THD) of $x\%$ when excited by a frequency ω_o, the scaled filter will also have a THD of $x\%$ when excited by a tone of the same amplitude, but a frequency of $\alpha\omega_o$. Extending this further,

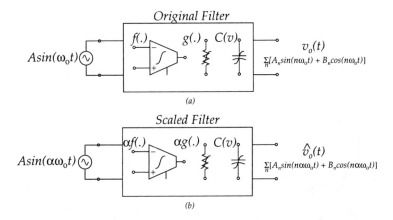

Figure 4.13. Distortion in weakly nonlinear filters.

if the original filter has a worst case THD of $x\%$ for an input tone of frequency ω_o and amplitude A, then the scaled filter will also have a worst case THD of $x\%$ when excited by a sinewave of amplitude A, but a frequency $\alpha\omega_o$. Therefore, **the worst case distortion of a scaled filter is independent of the scaling factor.**

In more general terms, since $\hat{v}_o(t) = v_o(\alpha t)$ for $\hat{v}_i(t) = v_i(\alpha t)$, we have in the frequency domain -

$$\hat{V}(f) = \frac{1}{\alpha} V(f/\alpha) \qquad (4.34)$$

The spectrum of $\hat{v}_o(t)$ is a frequency scaled version of the spectrum of $v_o(t)$. It therefore follows that if, for a multiple sinusoid input, $v_o(t)$ is expressed in terms of intermodulation or crossmodulation terms, the amplitudes of these components will remain the same for the scaled network.

To verify our conclusions on the distortion properties of scaled networks, simulations were run on a CMOS implementation of a constant-capacitance scaled Butterworth filter, whose bandwidth was programmable from $60 - 350\,\text{MHz}$. (The implementation and experimental results are described in the rest of this book, however that is not the focus here.) The filter bandwidth was set to $300\,\text{MHz}$, and the total harmonic distortion in the output waveform is plotted as a function of the input frequency. Then the bandwidth setting was changed to $174\,\text{MHz}$ and the above procedure was repeated again. The results are shown in Figure 4.14. In Figure 4.15, the x-axis is

88   HIGH FREQUENCY CONTINUOUS TIME FILTERS

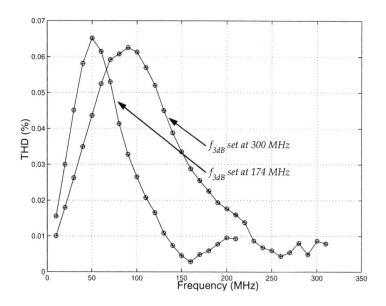

Figure 4.14.   Distortion simulations of a scaled Butterworth filter.

normalized with respect to the filter bandwidth setting for both the curves shown in Figure 4.14. As predicted by the theory, the two normalized distortion curves are in very good agreement. These are device-level simulations. The absolute distortion levels are not important because they depend on the accuracy of the transistor models used. What is of great significance is that the distortion curves are nearly identical when normalized to the bandwidth. To obtain accurate results, the simulator time-step was forcibly set in inverse proportion to the bandwidth. This ensures that the simulator converges to roughly the same degree at the end of a time step regardless of the filter bandwidth.

6.1   DYNAMIC RANGE IN SCALED FILTERS

Along with the noise properties of scaled networks discussed earlier in this chapter, the observation on distortion has important consequences in filter design. If the input signal to a filter is very small, the output is masked by the internal noise of the filter. If the input signal is very large, distortion effects set in. The dynamic range of the filter is defined as the ratio of the maximum rms output

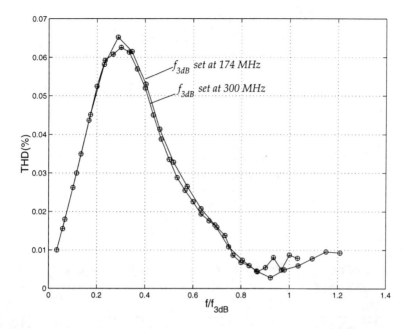

Figure 4.15. Distortion simulations of a scaled Butterworth filter, plotted with the x-axis normalized to bandwidth.

signal level permissible (with some acceptable distortion level) to the minimum output signal (usually the root mean squared output noise). If a constant-capacitance scaled network is used to implement a programmable filter, we have shown that the integrated output noise power is constant irrespective of the set bandwidth. From our discussion of time-scaling in nonlinear systems, we concluded that the maximum output signal level for a given level of distortion is independent of the set bandwidth too. From the above two statements, we conclude that the dynamic range of a constant-capacitance scaled network is independent of the scaling factor (assuming the effect of $1/f$ noise is negligible). Thus a constant-capacitance scaled filter represents a very desirable situation, summarized below :

- The relative shape of the frequency response of a scaled filter is maintained irrespective of the scaling factor.

- The output noise power of a scaled filter is independent of the scaling factor.

- The worst case distortion (and hence the maximal signal level) of a scaled filter is independent of the scaling factor.

- The dynamic range of a scaled filter is independent of the scaling factor.

Notice that while there are many scaling strategies to maintain frequency response, **only constant-capacitance** scaling keeps the dynamic range constant irrespective of the bandwidth. Hence no over-design is needed. If the original filter is designed optimally in terms of thermal noise and distortion, then any constant-capacitance scaled version of it will also be optimal, irrespective of frequency setting. This property is key to making possible the performance of the chip described in the next chapter.

7. IMPLEMENTATION OF SCALED INTEGRATORS IN CMOS VLSI

We now consider the practical issues that arise when scaled networks have to be implemented in CMOS VLSI. Although we have shown from other considerations that constant-capacitance scaling is the best technique, constant-conductance scaling is first considered for the sake of completeness.

7.1 CMOS IMPLEMENTATION OF CONSTANT CONDUCTANCE SCALING

In Figure 4.16 we show a constant-conductance scaled integrator. To scale the bandwidth of the circuit, the integrating capacitance can be discretely varied as shown in Figure 4.17. There are many problems with this approach. In high frequency designs, the capacitor values tend to be small. Splitting them into many smaller capacitors can result in poor matching. In order for the switches not to cause any degradation in the frequency response their on-resistance should be very low. This means that the switches would have to be large, and this results in large switch parasitic to ground (shown as C_p in Figure 4.17). However, this parasitic capacitance causes problems when the switch is off. Due to these limitations, this technique has been used mainly in filters operating at low frequencies [40] [41].

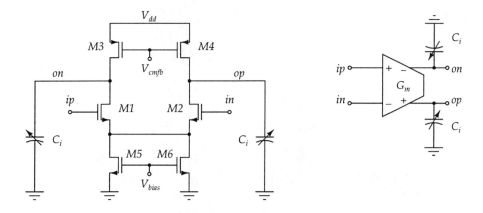

Figure 4.16. A constant-conductance scaled integrator.

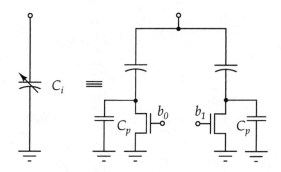

Figure 4.17. A digitally tuned variable capacitor.

7.2 CMOS IMPLEMENTATION OF CONSTANT CAPACITANCE SCALING

We now discuss a very efficient method of implementing constant-capacitance scaling in CMOS technology [35][42]. Consider the simplified schematic and symbol of a unit transconductance element shown in Figure 4.18. M1, M2, M3 and M4 are identical devices. M3, M4 are dummy devices used to maintain a constant input capacitance irrespective of the state of the switches. V_{cmfb} is a common-mode feedback signal, the generation of which is discussed in the next chapter. b and $\bar{b}$ are complementary switches. The small-signal equivalent circuits for differential excitation for the cases $b=1$ and $b=0$ are considered in Figure 4.19 (a) and (b) respectively. In (a), M1-M2 operate in strong inversion while M3-M4 are off. In (b),

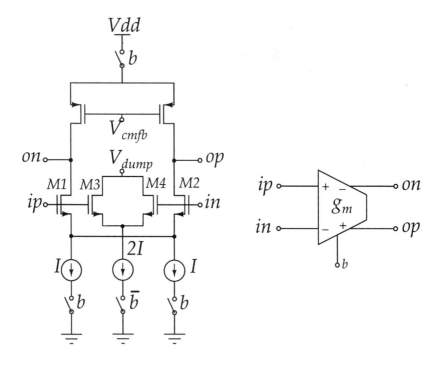

Figure 4.18. A unit transconductance element.

M3-M4 operate in strong inversion while M1-M2 are off. In **both** cases, **all** nodal capacitances in the network remain identical. This can be verified by using the complete model for the MOS transistor discussed in Appendix D. Notice that when the transconductor M1-M2 is turned off, **every** transconductance and conductance is reduced to zero, while **every** capacitance remains the same. Also, the unit transconductor does not contribute noise when it is turned off.

Now consider the parallel connection of two identical unit transconductors as shown in Figure 4.20. The table in the figure lists the various parameters of the composite transconductor (denoted by G_m, C_o etc.) as a function of those of the individual transconductors. Again, note that **every** node capacitance remains the same while **every** transconductance/conductance scales appropriately. The results can be extended to many unit transconductors connected in parallel. From this, we can conclude that any fully differential network created out of such transconductors and capacitors represents a constant-capacitance scalable system, with all the desirable properties outlined

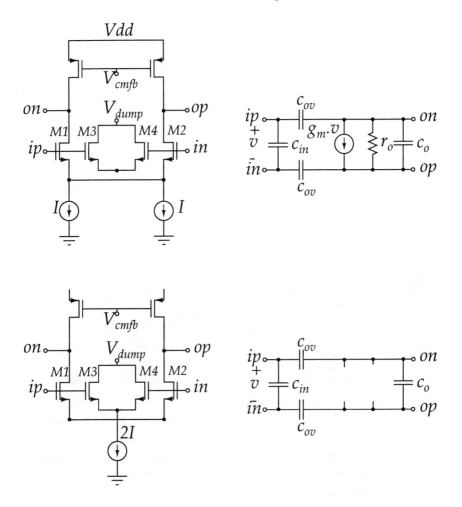

Figure 4.19. The small-signal equivalent circuit of the transconductor of Figure 4.18 under differential excitation, for the cases (a) $b = 1$ and (b) $b = 0$ (where b denotes the state of the corresponding switches in Figure 4.18).

in the previous section. **Notice that there are no switches in the signal path, and the scaling factor can be controlled by means of a digital word.** This is very convenient in modern communication systems, where a digital circuit controls the filter bandwidth.

8. CONCLUSIONS

In this chapter, we derived some properties of time scaled electrical networks. We have shown that constant-capacitance scaled filters are

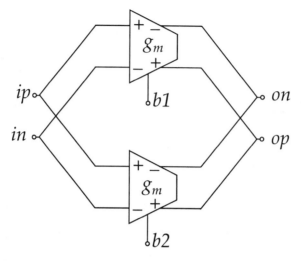

b1	b2	G_m	G_o	C_i	C_{ov}	C_o	DC Gain
0	1	g_m	g_o	$2c_i$	$2c_{ov}$	$2c_o$	g_m/g_o
1	1	$2g_m$	$2g_o$	$2c_i$	$2c_{ov}$	$2c_o$	$2g_m/2g_o = g_m/g_o$

Figure 4.20. Parallel connection of two unit transconductors.

optimal with respect to frequency response accuracy and dynamic range irrespective of the set bandwidth. The principle of a very efficient CMOS implementation of scaled filters has been presented. This will be the basis for the design of a 60 – 350 MHz Butterworth filter in a standard CMOS process. The detailed design is described in the next chapter.

Chapter 5

FILTER DESIGN

1. INTRODUCTION

To test the ideas presented in the previous chapter, a fourth order Butterworth filter with a maximum bandwidth of 350 MHz was designed in a 1.8/3.3 V 0.25 μm n-well digital CMOS process. This technology has low voltage devices (with an oxide thickness of 48 Å) with a nominal threshold of about 400 mV and high voltage devices (with an oxide thickness of 84 Å) with a nominal threshold of about 600 mV. The minimum drawn channel lengths for the low and high voltage devices are 0.3 μm and 0.45 μm respectively. The integrating elements used in the filter are suitably biased MOS accumulation structures. This makes the area of the capacitors a very small fraction of the area of the chip. In this chapter, we will discuss the design of the filter, address critical layout issues and present simulation results. The test strategy and measurement results are discussed in the next chapter. The top level block diagram of the test chip is shown in Figure 5.1. Each of these blocks will be described separately in the various sections to follow.

2. FILTER DESIGN

The fourth order Butterworth filter is realized as a cascade of two biquadratic sections. The center frequencies and quality factors of the biquads are shown in Table 5.1. The biquad with the higher quality factor is placed second in order to avoid complicated node voltage scaling procedures.

96 HIGH FREQUENCY CONTINUOUS TIME FILTERS

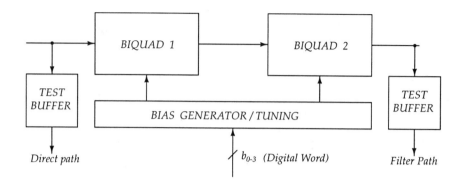

Figure 5.1. Block diagram of the test chip.

	Center Frequency	Quality Factor
Biquad 1	ω_o	0.5412
Biquad 2	ω_o	1.3066

Table 5.1. Center frequencies and quality factors of the individual biquads in a fourth order Butterworth filter with a bandwidth of ω_o.

The biquadratic sections used are shown in Figures 5.2 and 5.3. These are standard Gm-C biquad configurations based on the double integrator loop [43]. They only differ in the values of the damping transconductor. This particular choice results in reasonable values of capacitor spread. The transconductors are digitally programmable by a four-bit word. The design and layout considerations of the transconductor array are described in the next section. Since the integrating capacitors are in part due to layout and interconnect parasitics, only approximate values for the capacitors were put in during the schematic design phase. Final design centering of the filter response was done after extraction of the layout, by using a computer routine which calculated the explicit capacitor values required to be used.

Filter Design 97

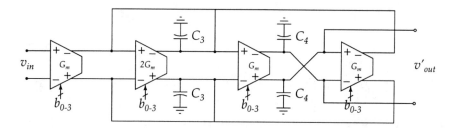

Figure 5.2. Biquad 1 used in the realization of the fourth order Butterworth filter prototype.

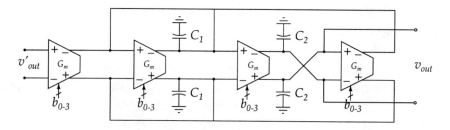

Figure 5.3. Biquad 2 used in the realization of the fourth order Butterworth filter prototype.

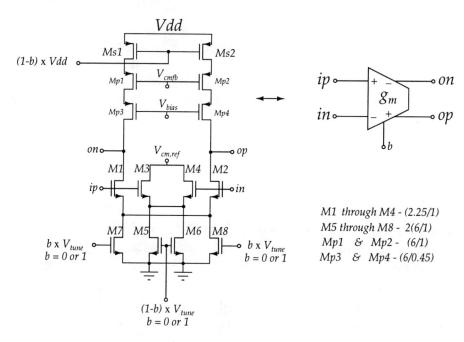

Figure 5.4. Unit transconductor cell.

98   HIGH FREQUENCY CONTINUOUS TIME FILTERS

3.   TRANSCONDUCTOR DESIGN

The schematic of the unit transconductance cell is shown in Figure 5.4. M1, M2, M3 and M4 are identical devices with M1-M2 forming the main differential pair of the transconductor, and M3-M4 forming a dummy differential pair. M7-M8 and M5-M6 form the current sources for the main and dummy differential pairs respectively. When M7-M8 are supposed to act as current sources ($b=1$), their gates are connected to a voltage V_{tune}, which is generated by a bandwidth tuning circuit (see Section 5). When these current sources are to be shut off ($b=0$), their gates are connected to ground and M5-M6 are connected to V_{tune}. A simple pass-transistor network is used to accomplish the switching of the gates of the current sources between V_{tune} and ground. This ensures that a differential pair (either the main one, M1-M2, or the dummy one, M3-M4) is always connected at the input, and thus the input capacitance is maintained constant independent of b. $V_{cm,ref}$ is a DC voltage equal to the common-mode DC level of the transconductor outputs (Section 4). Mp3 and Mp4 are cascode devices which enhance the output impedance of the PMOS current sources formed by Mp1 and Mp2. V_{cmfb} is a voltage derived from a common mode feedback circuit (see Section 4) that assures that the common mode level of on and op is equal to $V_{cm,ref}$. In our implementation, this value was chosen to be 1.2 V. Ms1 and Ms2 operate in the triode region and are used to switch the load current sources on and off.

The entire transconductor is made by connecting binary weighted unit cells in parallel, as shown in Figure 5.5. The low frequency gain of the transconductor is independent of the digital control word and is very nearly equal to the $g_m r_o$ product of NMOS devices. The digital control word for programming the transconductor is $b_0 b_1 b_2 b_3$. When the word is 0000, the cell transconductance is set to a value of $4 \times g_m$, where g_m is the transconductance of the unit cell. When the control word is 1111, the transconductance is $19 \times g_m$. The programming range of the transconductor is $19/4 = 4.75$.

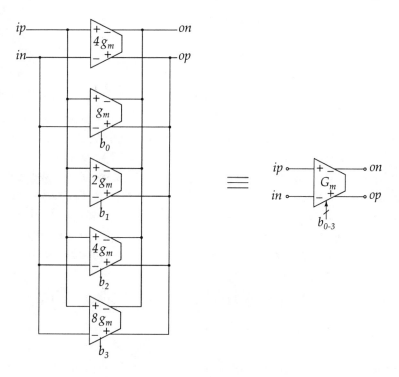

Figure 5.5.   Complete transconductor.

3.1 CHARACTERIZATION AND MODELING OF MISMATCH IN MOS TRANSISTORS AND ITS INFLUENCE ON DEVICE SIZING

Since the filter linearity requirements are modest, the transconductors are designed so that their linear range is limited. There exists the possibility that offsets generated within the filter will drive some transconductors into their nonlinear region, and result in a much poorer overall filter linearity than desired. Random device mismatches should be an important consideration while choosing the geometry of the devices.

In all matching studies, very simple models are assumed for the DC behavior of a MOSFET to keep the results manageable. This makes sense, because matching is determined by many factors which are not explicitly incorporated in the model parameters anyway. The equation used for the MOS transistor in saturation is [16]

$$I_{DS} = \frac{\frac{\mu_n C'_{ox}}{2a}}{1 + \theta_A (V_{GS} - V_T)} \left(\frac{W}{L}\right)(V_{GS} - V_T)^2 \quad (5.1)$$

The assumption that the transistor is strongly inverted is implied in the above model. The factor $\frac{\mu_n C'_{ox}}{\alpha}\left(\frac{W}{L}\right)$ is often called the "current factor" and denoted by β. For long channel MOSFETs channel length modulation can be neglected. The basic conclusion from all mismatch studies is that the matching between two transistors gets better if the individual devices have a large area. This is intuitively satisfying because a larger area means that the individual factors causing mismatch in the devices average out to a greater extent. Here, we will only summarize some important results of device matching studies [44] [45].

- **Device Geometry Dependence :** Threshold matching is given by

$$\sigma_{V_T} = \frac{A_{V_T}}{\sqrt{WL}} \quad (5.2)$$

where A_{V_T} is a function of the technology being used, and has a strong correlation to the thickness of the gate-oxide. For n-channel devices considered in [45], it has been empirically found that

$$A_{VT} = 3.4\,\text{mV}\mu\text{m} + 0.6\,\text{mV}\mu\text{m}\left(\frac{t_{ox}}{10\text{Å}}\right) \quad (5.3)$$

The reason for this dependence is not well understood. From the circuit design point of view, it suffices to say that A_{VT} gets smaller as technologies scale. The value of A_{VT} for p-channel devices is about 1.5 times that of the n-channel devices. This is because the n-well is counter-doped to tailor the threshold voltage of the PMOS transistors. To a first order the current factor mismatch is given by

$$\sigma_\beta = \beta \frac{A_\beta}{\sqrt{WL}} \quad (5.4)$$

A_β seems to decrease only slightly as feature size is reduced and a typical number to go by is about $0.02\mu\text{m}$. The current mismatch in two identically biased transistors can be shown to be

$$\frac{\sigma^2_{I_{DS}}}{I^2_{DS}} = \frac{4\sigma^2_{V_T}}{(V_{GS} - V_T)^2} + \frac{\sigma^2_\beta}{\beta^2} \quad (5.5)$$

For low values of gate overdrive, current mismatch is dominated by the mismatch in threshold voltages. All the above results are only valid for long channel devices. An accurate mismatch model for short-channel devices is yet to be discovered. Since we do not use short channel devices in critical signal paths in the filter design, this is not of concern.

- **Layout Considerations :** Apart from random offsets, offsets can manifest themselves due to bad layout . Strictly speaking, these are systematic in nature and can be avoided by paying attention to detail. Some examples of techniques to be adopted in the design to be described are - the use of dummy structures, liberal use of substrate contacts and avoiding routing metal over MOS gates at all costs [46]. A book [47] contains a useful collection of good layout practices.

The device sizes are chosen in such a way that random offsets due to threshold mismatches do not affect distortion severely. Consider the unit transconductance cell with offsets in the transistors as shown in Figure 5.6. The offset in the M3-M4 pair is not shown as its contribution to the input referred offset is negligible. The standard deviation of the input referred offset for a unit transconductance cell ($\sigma_{V_{unit,in}}$) is seen to be given by

$$\sigma^2_{V_{unit,in}} = \sigma^2_{V_{off,n}} + \left(\frac{g_{mp}}{g_{mn}}\right)^2 \sigma^2_{V_{off,p}} \qquad (5.6)$$

The value of $\frac{g_{mp}}{g_{mn}}$ was about 0.6. For the process used in the design, the oxide thicknesses for the n-channel and p-channel devices were 45 Å and 84 Å respectively. From (5.3) and (5.2) and the discussion of those equations

$$\sigma_{V_{off,n}} \approx 4.5\,\text{mV} \qquad (5.7)$$
$$\sigma_{V_{off,p}} \approx 8\,\text{mV} \qquad (5.8)$$

Using these values in (5.6), we get

$$\sigma_{V_{unit,in}} \approx 5.5\,\text{mV} \qquad (5.9)$$

When the digital control word is 0000, four unit cells are on, which effectively increases the area in (5.3) and (5.4). Hence, the standard

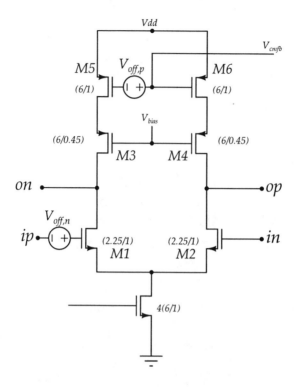

Figure 5.6. Offsets in a unit transconductance cell.

deviation of the input referred offset in the worst case is

$$\sigma_{V_{in},0000} = \frac{\sigma_{V_{unit.in}}}{\sqrt{4}} \approx 2.75\,\text{mV} \tag{5.10}$$

When the control word is set to 1111, all 19 unit cells are on. Their offsets average out to a greater extent and therefore we write

$$\sigma_{V_{in},1111} = \frac{\sigma_{V_{unit.in}}}{\sqrt{19}} \approx 1.25\,\text{mV} \tag{5.11}$$

Using the above numbers as a guide, Monte Carlo simulations were run on a device level implementation of the filter. A TI internal statistical circuit simulation tool called OCEAN was used to put in random offsets and run distortion tests. Not much meaning can be attached to the absolute values of distortion obtained from such simulations because they depend on the quality of the MOSFET models used. However much peace of mind was gained by observing that distortion levels did not change very much in the presence of

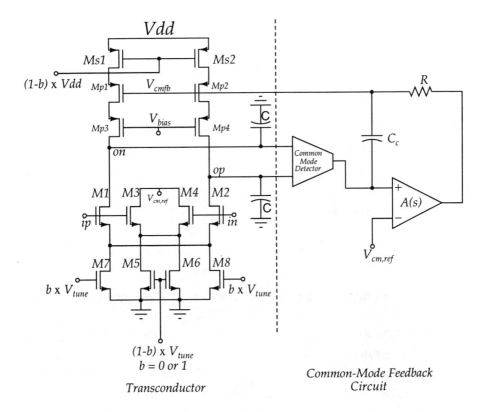

Figure 5.7. Block diagram of the common-mode feedback circuit.

offsets. Even putting in up to three times the expected offsets did not increase the levels of second harmonic distortion significantly. From this we conclude that the design is not "at the edge of a cliff" as far as mismatch is concerned. This also helps us to ascertain that the gate overdrive (about 350 mV, worst case) of the devices is sufficient to maintain the required filter linearity.

4. COMMON-MODE FEEDBACK CIRCUIT

Feedback loops are necessary to hold the common-mode voltage, at nodes operating fully differentially, at a constant value. The block schematic of the common mode feedback circuit, along with a unit transconductor whose output level is stabilized, is shown in Figure 5.7. The capacitors denoted by C are the integrating capacitors in the filter. Until further notice, assume that R and C_c are zero. The circuit functions as follows: The output common mode level is measured

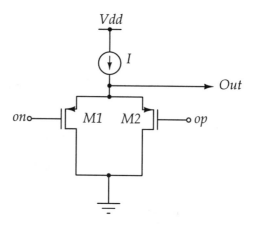

Figure 5.8. A differential pair as a common-mode detector.

by a common-mode detector and compared to the desired common-mode reference voltage ($V_{cm,ref}$) at the inputs of a high gain servo amplifier. The negative feedback loop is completed through Mp1 and Mp2. The grounded integrating capacitors are intended to stabilize the common mode feedback loop. In order that the common-mode level at the output of the transconductor be as close to $V_{cm,ref}$ as possible, the servo amplifier should be designed to have a large DC gain. This means that its bandwidth cannot be expected to be very high. If nothing were done, this would introduce significant phase shift at very high frequencies, which would in turn lead to instability in the common-mode loop. Even though the integrating capacitors provide the dominant pole in the loop, the delay through the servo amplifier might still not be acceptable. Hence, a "crossover network" formed by R and C_c is used to effectively remove the amplifier from the feedback loop at high frequencies. With this scheme, it is still possible to have a high DC gain, while maintaining loop stability.

Many choices exist for the common-mode detector. This circuit must be very linear; that is, it should not respond even to large differential signals. Two simple choices are shown in Figures 5.8 and 5.9. For the same bias current, the circuit of Figure 5.9 tends to be more linear. However, the advantage of the detector of Figure 5.8 is its capacitive input impedance, which can be absorbed into the integrating capacitors during the filter design. For a given bias current budget, both these detectors become more linear when the

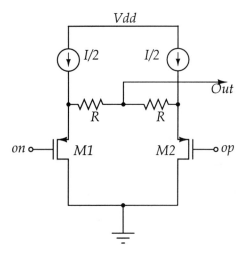

Figure 5.9. A degenerated differential pair as a common-mode detector.

gate overdrives increase. This means that for a given bias current, there is a trade off between the output impedance of the detector and its linearity.

A preliminary design of the common-mode feedback circuit, which employs the detector of Figure 5.8 is shown in Figure 5.10. Mc1 and Mc2 form the common-mode feedback detector. Their gates *on* and *op* are assumed to be connected to the outputs of the transconductor, as shown in Figure 5.10. Mc3, Mc4, Mc11 and Mc12 form the servo amplifier. The voltages *biasn* and *biasp* are generated by a bias generation circuit, discussed in Section 4.1. The common-mode reference voltage ($V_{cm,ref}$) is level-shifted by Mc5, so as to be compatible with the levels generated by the detector. Mcs1 and Mcs2 are PMOS devices operating in the triode region. They match the small voltage drop across the switches Ms1 and Ms2 in the transconductor (see Figure 5.4). Mc9 is a *p*-channel device acting as a resistor, which along with C_c takes the servo amplifier out of the feedback loop at high frequencies. This design, however, has a problem. If high linearity is desired in the common-mode detector, its gate overdrive voltage has to be increased, resulting in a higher output impedance for the detector. Thus, the detector would take a long time to charge any capacitance at the node V_{cmfb}, leading to excess delay and possible instability. One solution to this problem is

106   HIGH FREQUENCY CONTINUOUS TIME FILTERS

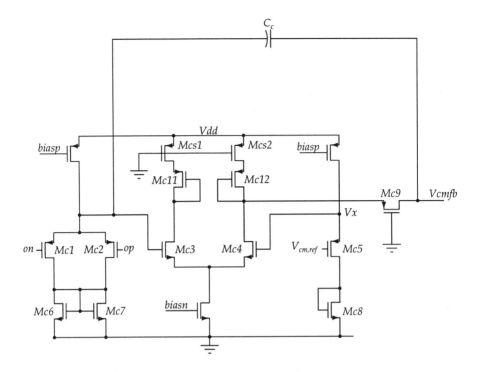

Figure 5.10.   Preliminary design of the CMFB loop.

to increase the sizes of the M1 and M2, while maintaining the same current density in them. The disadvantage of this approach is that the common-mode detector now presents a large capacitive load to the filter.

Figure 5.11 shows the final CMFB loop, which avoids the problem discussed above. A transconductor used in the filter is shown towards the right of the figure. The cascode devices for the current sources in the transconductor have been omitted for simplicity. The tail current for the transconductor is derived from a bandwidth tuning circuit (see Section 5). The voltages *biasn* and *biasp* in the CMFB circuit are derived from a bias generator circuit, which is described in Section 4.1. An *n*-channel source-follower Mc10 is used to buffer the output of the common-mode detector. Since the follower is biased to have a very low output impedance, it easily drives any capacitance at the node V_{cmfb}. C_c is a MOS accumulation capacitor. The way it is connected automatically provides a comfortable bias voltage pushing it deep into accumulation.

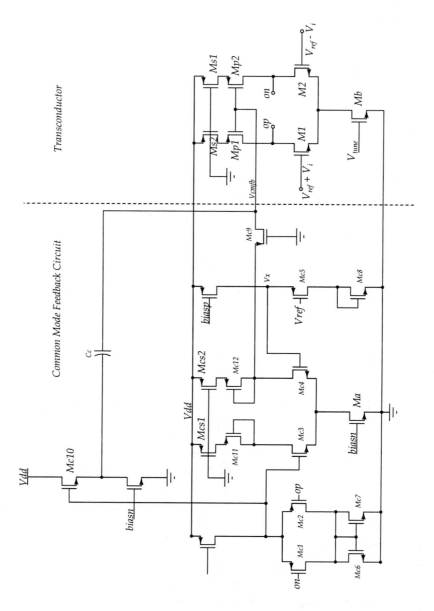

Figure 5.11. Final CMFB loop.

Mc1 and Mc2 are low voltage devices. To ensure that they do not have a source-drain voltage drop of more than 1.8 V, the diode connected devices Mc6 and Mc7 are added in series with the drain.

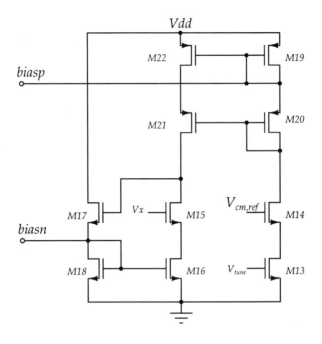

Figure 5.12. Schematic of the bias generation circuit for the CMFB loops.

4.1 BIAS GENERATION FOR THE COMMON-MODE FEEDBACK CIRCUITRY

Bias currents for the common-mode feedback circuit are generated in such a way as to eliminate/minimize any systematic offset in the output common-mode level of the transconductors. This is particularly important in this design because the integrating capacitors are bias dependent. Consider the circuit of Figure 5.11. When the transconductor output common-mode level is exactly equal to the desired voltage $V_{cm,ref}$, the current density in Mc12 must be identical to that in Mp1 and Mp2 *if there is to be no systematic offset in the loop.* This means that Ma and Mb must have the same current density. This cannot be achieved if Ma is biased with $V_{GS} = V_{tune}$ (recall that V_{tune} is generated by the bandwidth tuning circuit to be described in Section 5) because of the difference in the drain potentials of Mb and Ma. The circuit of Figure 5.12 is used to compensate for the change in current density due to this drain voltage difference. In essence, it generates an appropriate voltage, which when applied to the gate of Ma, would ensure that the current densities in Ma and Mb are equal.

V_x in Figure 5.12 is connected to the source node Mc5 in Figure 5.11. M14 and M13 are sized so that they have the same current densities as M1 and Mb in Figure 5.11 respectively. M13, M16, M17 and M18 are identical devices. M15 is twice as wide as Mc3 and M16 is identical to Ma. M19 through M22 form a current mirror. The circuit functions as follows: the current generated by M13 is mirrored through the *p*-channel current mirror. The negative feedback loop formed by M15, M16, M17 and M18 forces the current in M16 to be that in M13. This voltage, denoted as *biasn* is used to as the gate drive for Ma in the CMFB circuit. This negative feedback arrangement assures that Ma and Mb in Figure 5.11 maintain the same current densities over process and temperature, thereby minimizing systematic offsets in the common-mode loop. The potential at the gate of M19 (*biasp*) is used for all the *p*-channel current sources in the common-mode feedback circuit.

5. FREQUENCY TUNING SYSTEM

In order to keep the filter bandwidth relatively constant with temperature and process variations, the transconductance values of all the differential pairs used in the filter are servoed to a stable external resistor. This is a low-complexity scheme without the feedthrough problems associated with more complicated tuning techniques. The conceptual schematic of a conventional resistor servo tuning loop is shown in Figure 5.13. A voltage V_{in} is applied to the transconductor, and its output current is compared to the current generated by applying V_{in} to a stable external resistor. The difference in currents is integrated with the opamp and the capacitor. Due to negative feedback, the output of the opamp settles to a value such that $G_m = \frac{1}{R}$. The voltage developed is used to control other transconductors on the chip.

Rather than use such an explicit DC loop to accomplish a "resistor-servo", we exploit the square law characteristics of the MOSFET. The circuit will be developed in this section. The following model is assumed for the MOS transistor operating in strong inversion and saturation [16].

$$I_{DS} = \frac{\mu_n C'_{ox}}{2a} \left(\frac{W}{L}\right)(V_{GS} - V_T)^2 \qquad (5.12)$$

110   HIGH FREQUENCY CONTINUOUS TIME FILTERS

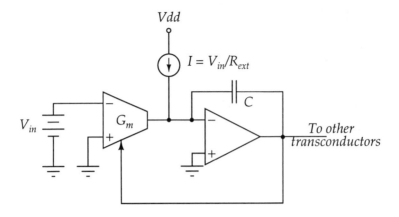

Figure 5.13.   Conceptual schematic of a conventional resistor-servo loop.

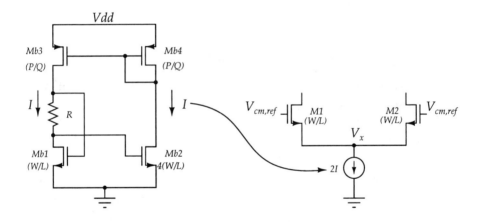

Figure 5.14.   Basic fixed transconductance bias circuit.

where all the symbols have their usual meanings. Mobility reduction due to gate field is neglected; this effect will be considered separately later on. Since long channel transistors are used, velocity saturation effects and finite drain conductance can be neglected. The basis for all the fixed-transconductance-bias circuits in this section is shown in Figure 5.14 [48]. The current generated in the bias cell is used as the tail current for the differential pair transconductor formed by M1-M2 shown towards the right of the figure. In the quiescent state, the gates of M1 and M2 are at a potential $V_{cm,ref}$, the common-mode reference of the filter. (This condition is forced by the common-mode feedback circuitry for every transconductor output in the filter, and one such output is assumed to be driving the gates of M1 and M2). The current mirror formed by M3b and Mb4 forces identical currents through Mb1 and Mb2. We have

$$I_{Mb1} = \frac{\mu_n C'_{ox}}{2a} \left(\frac{W}{L}\right)(V_{GSb1} - V_T)^2 \tag{5.13}$$

$$I_{Mb2} = \frac{\mu_n C'_{ox}}{2a} \left(\frac{4W}{L}\right)(V_{GSb2} - V_T)^2 \tag{5.14}$$

$$I_{Mb1} = I_{Mb2} = I \tag{5.15}$$

$$V_{GSb1} - V_{GSb2} = IR \tag{5.16}$$

From (5.13), (5.14) and (5.15), we get

$$V_{GSb1} - V_T = 2(V_{GSb2} - V_T) \tag{5.17}$$

From (5.16) and (5.17) we obtain

$$V_{GSb1} - V_T = 2IR \tag{5.18}$$

and, since $g_m = dI/dV_{GS}$,

$$\boxed{g_m|_{Mb1} = \left(\frac{2I}{V_{GSb1} - V_T}\right) = \frac{1}{R}} \tag{5.19}$$

The circuit stabilizes to a state where the current is such that the transconductance of Mb1 is maintained at $1/R$, irrespective of V_T, μ or temperature. However, the source potentials of M1-M2 are at V_x, while Mb1 and Mb2 have their sources grounded, which means the threshold voltages of Mb1-Mb2 are not the same as those of M1-M2. To remedy the situation we resort to the scheme shown in Figure 5.15. Here, the voltage of the source node of M1-M2 is sensed and the sources of Mb1 and Mb2 are "pushed up" by the same amount. Now Mb1, Mb2, M1 and M2 have the same threshold voltages and a better performance can be expected.

Implementation as shown in Figure 5.15 would place extreme demands on the output resistance of the "battery" used to push the sources of Mb1 and Mb2 to V_x. A better way of doing this is by the arrangement shown in Figure 5.16, where we exploit the fact

112   HIGH FREQUENCY CONTINUOUS TIME FILTERS

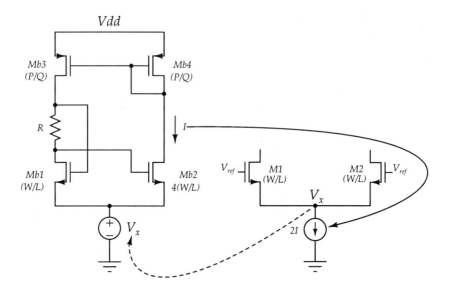

Figure 5.15.   Deriving an improved fixed transconductance bias circuit.

that a current $2I$ would flow through the battery. If that current was sunk through a current sink, a very small current (ideally zero) would have to flow through the battery, and a much higher output resistance could be tolerated. Using the modification suggested in Figure 5.16 in Figure 5.15, we get the circuit of Figure 5.17. Mb5 is used to mirror I flowing through Mb4. This current is multiplied by 3 in the NMOS mirror formed by Mb8 and Mb9. Mb9 thus acts as a current sink of value $3I$. Of this, $2I$ is supplied by Mb1 and Mb2, and I flows through Mb6 (which is of the same dimensions as M1 in Figure 5.15). Clearly, the voltage at the source of Mb6 is V_x, just as in Figure 5.15.

The complete "fixed transconductance bias" circuit, along with a transconductor used in the filter, is shown in Figure 5.18. The output impedances of the PMOS mirrors have been enhanced with cascode devices, and the drain of Mb6 has been tied to the gate (this is done because Mb6 then operates *exactly* in the same manner as the input devices in the differential pair). Mb7, Mb10 and Mb11 are added so that the mirror devices Mb8 and Mb9 can have the same drain-source voltages for a wide range of $V_{cm,ref}$. In fact, the circuit would work well even if Mb8 and Mb9 were operating in the triode region, corresponding to a very low value of $V_{cm,ref}$. C_c is a large valued

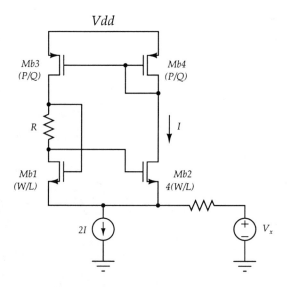

Figure 5.16. Scheme with reduced sensitivity to the output resistance of V_x in Figure 5.15.

MOS accumulation capacitor which feeds forward across Mb10 at high frequencies, improving the stability of the loop formed by Mb7, Mb8, Mb10 and Mb11. The resistor R_b isolates the bias voltage generated at the gate of Mb8 from the capacitive load the circuit sees and increases the effectiveness of C_c at high frequencies.

Figure 5.19 shows the simulated variation of the transconductance of a differential pair whose tail current is derived from the circuit of Figure 5.18. R was set to 300 Ω. Notice that the circuit performs fairly well over a hundred degree temperature range in spite of using the (crude) square law model for the devices.

We will now investigate the effects of mobility reduction due to gate field. Since the length of the channel is chosen to be large, short channel effects and channel length modulation can be neglected. To first order, the effective mobility of the carriers in the channel can be written as [16]

$$\mu_{eff} = \frac{\mu_n}{1 + \theta_A(V_{GS} - V_T) + \theta_B V_{SB}} \qquad (5.20)$$

and hence

$$I_{DS} = \frac{\frac{\mu_n C'_{ox}}{2a}\left(\frac{W}{L}\right)(V_{GS} - V_T)^2}{1 + \theta_A(V_{GS} - V_T) + \theta_B V_{SB}} \qquad (5.21)$$

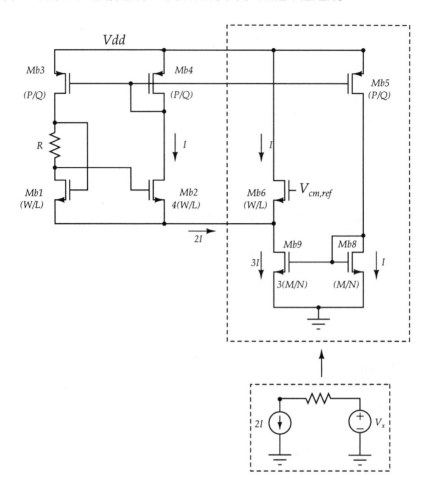

Figure 5.17. Circuit implementation of the scheme of Figure 5.16.

For simplicity we assume that V_{SB} is zero. Then,

$$I_{DS} = \frac{\frac{\mu_n C'_{ox}}{2a}\left(\frac{W}{L}\right)(V_{GS} - V_T)^2}{1 + \theta_A(V_{GS} - V_T)} \qquad (5.22)$$

Solving the above equation for $V_{GS} - V_T$, we get

$$V_{GS} - V_T = \frac{\theta_A a I_{DS}}{\mu_n C'_{ox}\left(\frac{W}{L}\right)} + \sqrt{\frac{2a I_{DS}}{\mu_n C'_{ox}\left(\frac{W}{L}\right)} + \frac{a^2 \theta_A^2 I_{DS}^2}{\left[\mu_n C'_{ox}\left(\frac{W}{L}\right)\right]^2}} \qquad (5.23)$$

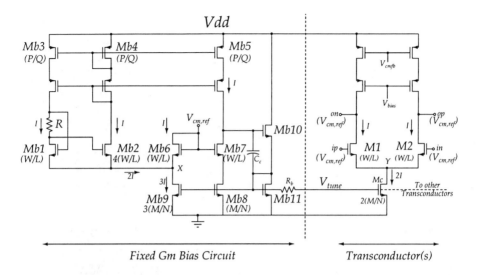

Figure 5.18. Complete "fixed transconductance bias" circuit, shown along with a transconductor used in the filter.

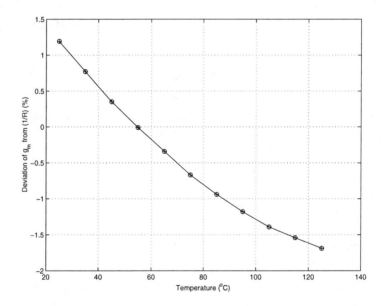

Figure 5.19. Percent g_m variation of the differential pair with temperature.

If the second term under the square root is neglected, we can rewrite (5.23) as

$$V_{GS} - V_T = \frac{\theta_A a I_{DS}}{\mu_n C'_{ox} \left(\frac{W}{L}\right)} + \sqrt{\frac{2a I_{DS}}{\mu_n C'_{ox} \left(\frac{W}{L}\right)}} \qquad (5.24)$$

116 HIGH FREQUENCY CONTINUOUS TIME FILTERS

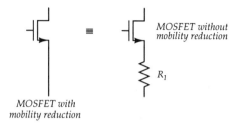

Figure 5.20. Equivalent representation of mobility reduction.

Comparing with (5.13), this means that, at least for small values of θ_A, reduced mobility is equivalent to having a resistor of value $R_1 = \frac{\theta_A a}{\mu_n C'_{ox}\left(\frac{W}{L}\right)}$ in series with the source of a MOSFET without mobility reduction, as illustrated in Figure 5.20. Therefore, the effect of mobility reduction is to modify the value of the transconductance setting resistor R. In order to get a rough idea of the value of θ_A and V_T, simple tests on the MOS model used in simulation can be run. Figure 5.21 shows the drain current of a long channel MOSFET (in the technology used in this design) biased with a small drain-source voltage ($V_{DS} = 100\,\text{mV}$). Notice that the curve is almost a straight line in strong inversion, suggesting that θ_A is very small. Figure 5.22 shows the transconductance of the device. From the data, the value of θ_A was calculated to be about $(9\,\text{V})^{-1}$, giving $\theta_A(V_{GS} - V_T) \approx 0.04 << 1$. We can conclude, therefore, that the deviation of the transistors from square law behavior is very small in the range of voltages that we are interested in. In this light, the results shown in Figure 5.19 seem reasonable.

6. PARASITIC CAPACITANCES

In this section we briefly discuss various issues associated with the parasitic capacitors used in the filter design. The explicit integrating capacitors are MOS accumulation devices. However, since the pole frequencies are very high, we cannot avoid interconnect and depletion parasitics accounting for a portion of the integrating capacitance. We do not discuss MOS capacitors here because they have been thoroughly analyzed in Chapter 2. We focus on depletion and interconnect capacitances.

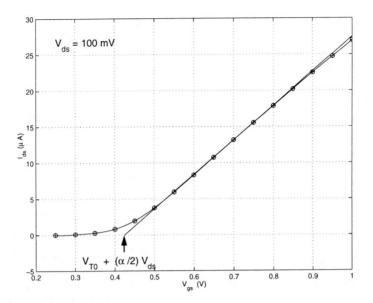

Figure 5.21. I_D as a function of V_{GS}.

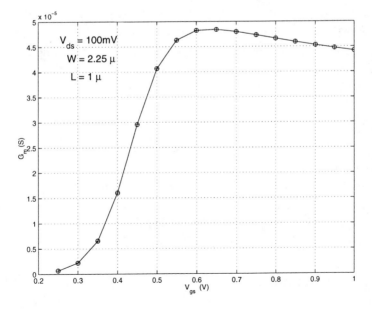

Figure 5.22. g_m as a function of V_{GS}.

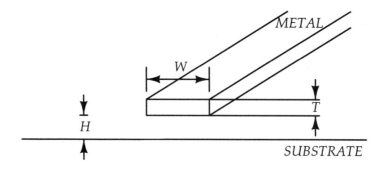

Figure 5.23.   Cross section of an interconnect line.

6.1   DEPLETION CAPACITORS

These are the (bias dependent) drain to bulk capacitances that occur at the output nodes of each transconductor. These are incorporated into the MOSFET models used for circuit simulation. Since we are considering filters in the very high frequency range, attention needs to paid to the parasitic series resistance of the depletion capacitors which depends strongly on the filter layout. If the equivalent series resistance of the depletion capacitor is very large, it will cause undesirable phase-shifts in the integrators leading to an incorrect filter response. Hence, the layout was liberally sprinkled with substrate contacts.

6.2   INTERCONNECT CAPACITORS

Parasitic interconnect capacitance forms a part of the integrating capacitance and must be taken into account during the filter design. In this section, we examine the nature of interconnect capacitance and some layout considerations which increase the quality factor of this capacitance from what might otherwise be achieved. We also present a layout technique which keeps the interconnect capacitances balanced.

Consider the interconnect line shown in Figure 5.23. Since an analytical expression for the capacitance of the line per unit length is extremely complicated giving us little insight, we will use a simpler but empirical formula. For $0.3 < (W/H), (T/H) < 30$, it has been

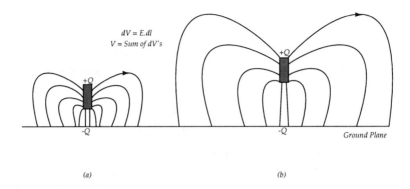

Figure 5.24. Comparing the capacitance of a thin interconnect line for two different heights over a ground plane.

found that [49]

$$C = \epsilon_r \left[1.15 \left(\frac{W}{H}\right) + 2.8 \left(\frac{T}{H}\right)^{0.222} \right] \quad (5.25)$$

is within 6% of the analytically calculated value. In this design, the value of W/H used was about 1. Typical values for T/H range from about $0.25 - 1.5$ depending on the layer of metal used for the interconnect. Notice that for a given value of W/H, the value of C depends on the value of $\left(\frac{T}{H}\right)^{0.222}$ and is therefore very insensitive to T/H. Hence, for small W/H, the capacitance of the line does not decrease significantly even if the interconnect is located on the topmost metal layer in the process. This is in sharp contrast to what is predicted by the parallel plate approximation. This discrepancy is because most of the capacitance of a small width interconnect line is due to fringing. More intuition about this is given next. Consider the capacitance of two thin interconnect lines over a ground plane as shown in Figure 5.24. Positive charges of value Q coloumbs are placed on each conductor. The potential of the conductor with respect to the ground plane needs to be obtained in order to determine its capacitance. To determine the potential difference, we travel along a field line and sum the infinitesimal quantities **E.dl** along the line.

In Figure 5.24(a), the density of electric field lines (and the magnitude of electric field) is very high while the lines are short in length. In Figure 5.24(b), the density of electric field lines is smaller while the lines are longer. Because of this, the value of the potential difference (and hence, the capacitance) between the conductor and the ground plane obtained for the two cases is close to each other. **The capacitance of a thin metal interconnect line is very insensitive to the thickness of the field oxide, and to the width of the line itself.**

In the present design, the second and third metal levels (called metal2 and metal3) were used as interconnect. A ground shield of metal1 was used below long lines. This is done for two reasons. First, it reduces signal coupling to the substrate. Second, the parasitic capacitance has a well defined path to ground, rather than an unknown (resistive) path through the substrate. The shield would increase the capacitance of the lines by a small amount, but it is certainly worth it in a very high frequency design such as the one undertaken here.

To ensure that the interconnect parasitics on balanced circuit nodes are equal, a layout technique which is "balanced by construction" is used. This is illustrated in Figure 5.25(b). When two pairs of balanced circuit nodes are to be connected, in the layout each of the lines is extended well beyond the junction. This strategy pays off very well in the time saved in trying to fix problems with circuit balance upon extraction of the circuit schematic from the layout. In our design, the layout-extracted netlist was balanced to more than 99.99% for frequencies up to 1 GHz.

7. BIQUAD LEVEL LAYOUT

In this section we discuss the filter layout at the biquad level. We will present a layout technique which cancels (to first order) systematic offsets due to small uniform gradients in gate oxide thickness. Consider the transconductor used as the building block of the design. The schematic and the layout are shown in Figure 5.26. The dummy n-channel devices switched in when the main portion of the transconductor is off are not shown in the schematic to avoid clutter in the diagram. In the layout they are represented by NL' and NR'. *ip-in* and *op-on* represent the inputs and outputs of the transconductor respectively. Notice carefully the correspondence between the schematic and layout. For example, the connection of *on* to *PL* and *NL* in the schematic is shown in the layout by a

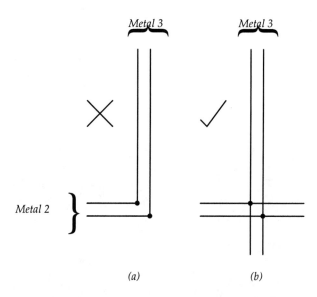

Figure 5.25. An inherently balanced interconnect technique.

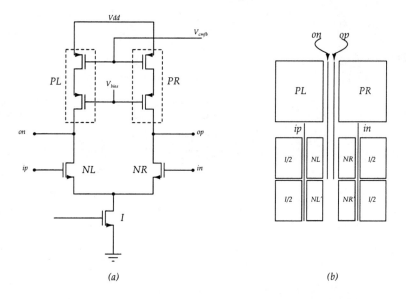

Figure 5.26. Schematic and layout of the transconductor used in the filter.

vertical line close to PL and NL. When the transconductor layout is mirrored about a vertical axis, the relative positions of PL and PR are interchanged. The same holds for NL and NR. In the case of a uniform oxide, the mirroring of the layout will not change the

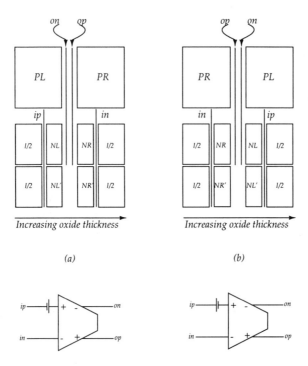

Figure 5.27. Two transconductor layout representations, and their equivalent circuits in the presence of an oxide gradient.

electrical equivalent circuit of the transconductor. In the presence of a linear gradient in oxide thickness, however, the two layouts are not identical. To first order, the effect of a small gradient can be represented as an equivalent DC offset voltage at the input of the transconductor. Depending on the choice made in choosing the layout, the input referred offset due to oxide gradient can be either positive or negative. This is shown in Figure 5.27. At the biquad level, this systematic offset can be reduced by a large factor (and in some cases even be cancelled). For purposes of illustration, consider the biquadratic section shown in Figure 5.28. Assume that all the transconductors are nominally identical, and have systematic offsets as shown. For the total output offset it can be shown that

$$V_{off,biquad} = V_{off1} + V_{off4} + V_{off2} - V_{off3} \quad (5.26)$$

From Figure 5.27, the sign of the systematic offset due to an oxide gradient can be changed depending on which side is associated with

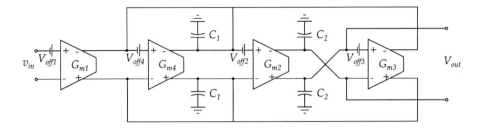

Figure 5.28. Schematic of a biquadratic section.

NL and PL. Clearly, the way the transconductors are connected makes a difference in the total systematic output offset of the biquadratic section. First, assume that all the transconductors are such that the input in the left-side is associated with NL. This corresponds to the layout of Figure 5.27(a). Then, the offsets of all the transconductors are the same in magnitude and sign. In this case, we observe from (5.26) that V_{off2} and V_{off3} cancel each other, while V_{off1} and V_{off4} add, resulting in a residual systematic offset. To reduce this to zero, we need to change the sign of V_{off4}. This is accomplished by associating the right-side of G_{m4} with NL (this corresponds to the layout of Figure 5.27(b)). The resulting biquad layout, shown in Figure 5.29, is to first order compensated for a small linear gradient in the oxide thickness along the direction shown. Notice that this does not cost anything, while it most likely improves performance. If care were not taken, one could end up with all the systematic offsets adding, and this could cause circuit problems.

The reader may recognize that the proposed layout technique is some sense similar to the common-centroid layout procedure used to cancel process and temperature gradients. Employing a common centroid geometry at the device level in a very high frequency design such as this would increase interconnect parasitics, and might even be harmful to MOS transistor matching due to metal coverage effects [46]. The method we adopt accomplishes the same purpose, and comes for free.

8. SIMULATION RESULTS

In this section we present post-layout simulation results for the filter design. These simulations were run at 65°C.

124   HIGH FREQUENCY CONTINUOUS TIME FILTERS

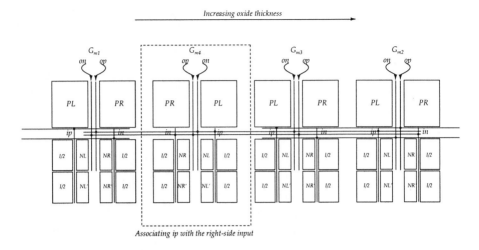

Figure 5.29. Layout of a biquadratic section tolerant to small oxide gradients.

8.1 AC SIMULATIONS

Figure 5.30 shows the response of the filter as the frequency control word is stepped from 0000 thru 1111. The external resistor setting the transconductance of the differential pairs used in the filter was such that the 3 dB bandwidth is 300 MHz when the control word is set to 1111. A linear scale is used in Figure 5.30 to show that the response scales very well even at such high pole frequencies and over a tuning range of almost 5x. The response is plotted on a logarithmic scale in Figure 5.31. Notice the absence of any peaking in response at very high frequencies. This can be attributed to the very simple nature of the transconductor (without any internal nodes) used. Figure 5.32 shows the response at the bandpass output of the first biquad. The peaks in the response are almost perfectly identical, as they should be.

The filter response as a function of temperature is shown in Figure 5.33. For this simulation the frequency control word was set at 1111. As is evident from the plot, the "fixed transconductance bias" tuning loop does a reasonable job. The variation in the 3 dB bandwidth over a 100 degree range is about ±2%. This is more than adequate for the intended application.

Figure 5.34 shows the balance of the lowpass and bandpass outputs of each of the biquads. What is meant by balance is the following.

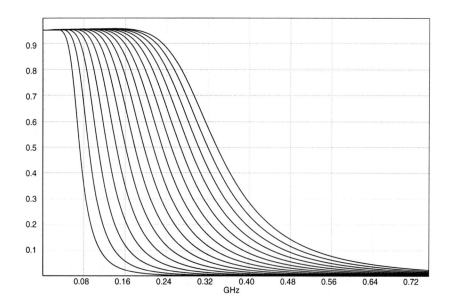

Figure 5.30. Frequency response as a function of programming word – linear scale.

If two nodes operate in a fully differential fashion, their magnitude responses must be identical. A measure of the "unbalance" of the layout is a ratio (in decibels) of the magnitude responses of the single ended outputs. From Figure 5.34 we notice that the worst case imbalance is about 5 millidB (1.0005) up to 1 GHz. These balance plots are for the raw layout - no "tweaking" of the layout is necessary (commonly achieved by adding stubs of interconnect at various points in the layout) because of the "balanced by construction" approach adopted in the layout of the chip. Our approach avoids the anxiety and frustration caused by noticing imbalances which cannot be easily traced to their root causes.

8.2 NOISE SIMULATIONS

Figure 5.35 shows the square root of the power spectral density of the filter output as the frequency control word is stepped from 0000 thru 1111. For this simulation the $1/f$ noise sources in the transistor models were turned off in order to demonstrate the noise properties

126   HIGH FREQUENCY CONTINUOUS TIME FILTERS

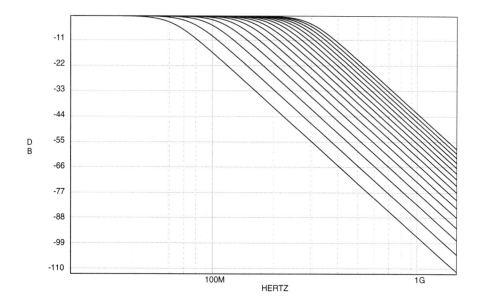

Figure 5.31. Frequency response as a function of programming word – logarithmic scale.

of frequency scaled networks discussed in Chapter 4. Figure 5.36 shows the plots of integrated output noise for various frequency settings. What is plotted is

$$v_{no}(f) = \sqrt{\int_0^f S_n(u)\,du} \qquad (5.27)$$

where $S_N(f)$ is the output noise power spectral density of the filter. As discussed in the previous chapter, the plots demonstrate the fact that the integrated output noise stays constant irrespective of frequency setting.

8.3   DISTORTION SIMULATIONS

Figure 5.37 shows the distortion components in the filter output when random offsets are incorporated into the circuit. For this, OCEAN, an internal statistical circuit simulation program, was used. The input was a 200 mV$_{pp}$ sinewave at 100 MHz. The filter corner was set to 300 MHz. The reason for the choice of a 100 MHz tone is the

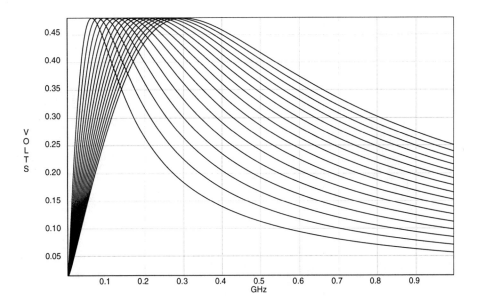

Figure 5.32. Bandpass output of Biquad 1 as a function of programming word.

following. If a very low frequency were used, the currents flowing through the integrating capacitors would be negligible. The nonlinearities of the transconductors would compensate for one another and a very low distortion would be obtained. For example, for DC, the input-output curve can be shown to be a perfect straight line. On the other hand, putting a tone close to the passband edge is also not correct because all the distortion components (the most dominant one being the third harmonic) would lie well outside the passband of the filter, and be partially filtered out. A good choice then, is to use a frequency such that the third harmonic of the input lies around the passband edge of the filter. Since the filter corner was set to 300 MHz, the input frequency was chosen as 300 MHz/3 = 100 MHz. The distortion simulations in Chapter 4 also lend credence to our observations here. Random normally distributed offset voltages with a standard deviation of 3 mV were used at the input of every transconductor. From the discussion on threshold mismatch in MOS devices, observe that this corresponds to about three times more offset than expected. In spite of this, the levels of

128 HIGH FREQUENCY CONTINUOUS TIME FILTERS

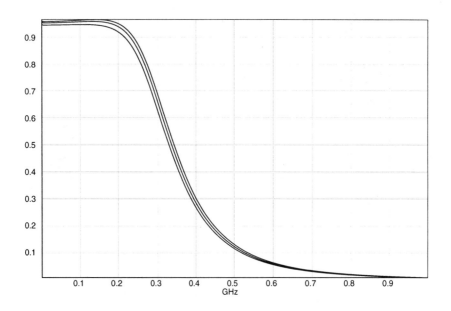

Figure 5.33. Filter response for temperature values of 25, 65 and 125°C.

second harmonic distortion are lower than the third harmonic, and there does not seem to be much variability in the total harmonic distortion. We conclude therefore that the design is not at the "edge of a cliff."

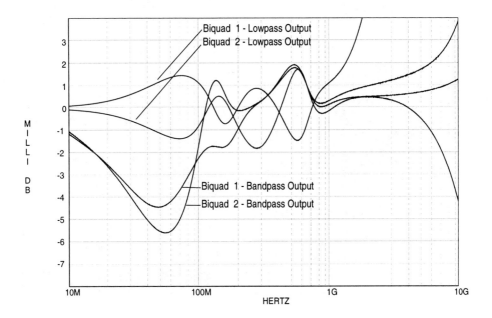

Figure 5.34. Balance in the filter layout.

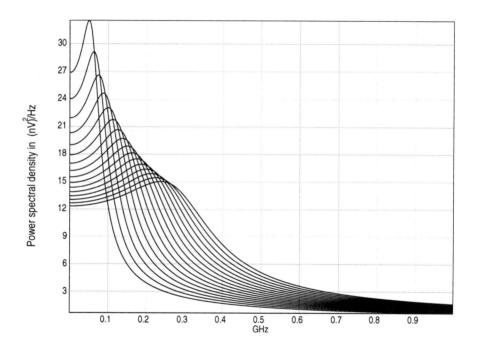

Figure 5.35. Output noise power spectral density plots as the frequency control word is stepped from 0000 thru 1111.

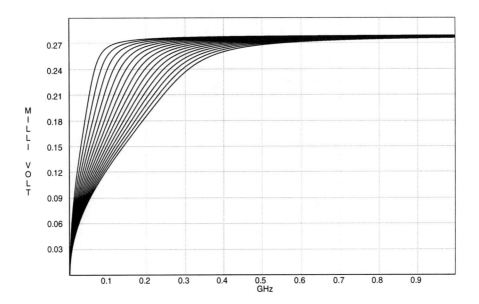

Figure 5.36. Integrated output noise as the frequency control word is stepped from 0000 thru 1111.

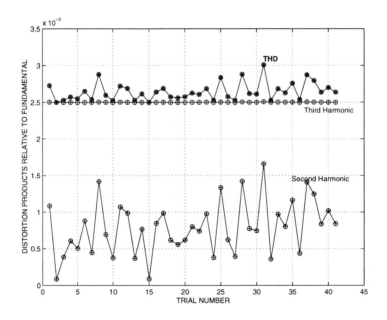

Figure 5.37. Results of Monte Carlo simulations of the filter with random offsets. The largest THD is 0.3%.

Chapter 6

FILTER TESTING AND MEASURED RESULTS

1. INTRODUCTION

In this chapter we discuss the measurement strategy and test results of a prototype filter fabricated in a $0.25\,\mu$m n-well digital CMOS technology at Texas Instruments. The detailed design of the filter was discussed in the previous chapters. Measurement of filter characteristics at very high frequencies requires special precautions in the design of the board and the pinout of the integrated circuit. For input frequencies that lie in the filter stopband, the output is very small in magnitude. Even small amounts of feedthrough from the input can result result in gross measurement errors. To reduce direct feedthrough problems as much as possible, the input and output pins of the filter chip must be a far away from each other.

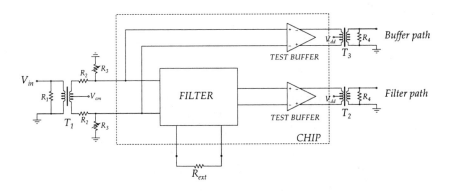

Figure 6.1. Simplified schematic of test setup.

A block schematic for the filter test set up is shown in Figure 6.1. T_1 converts the single ended input signal into a fully differential form. R_1 and R_4 are 50 Ω resistors which act as terminations for the measurement equipment. R_2 and R_3 have two functions. First, they dampen parasitic resonances caused by the filter input impedance in combination with package inductance. Second, by varying the two R_3's differently one can add controlled amounts of differential offset to the filter input. (It turns out that the AC imbalance caused by this does not create problems for the values used.) The purpose of this feature is to be able to study distortion performance of the filter in the presence of a DC offset. R_2 was set to 50 Ω and R_3 was made variable in the range 800 − 1000 Ω. The test buffers are designed as to have a current output. T_2 and T_3 convert the differential component of the output current into a single-ended voltage.

Under some very unrestrictive conditions, it is possible with this setup to measure the frequency response of the filter very accurately [8] in spite of the non-idealities of the test buffers and transformers. In our measurements, we make the following assumptions

- Both test buffers are matched.

- The signal paths through T_2 and T_3 are matched.

- The test buffers and the filter are unilateral, i.e they have no reverse transmission.

- The input impedance of the test buffers is purely capacitive, and has been accounted for in the filter synthesis.

From Figure 6.2, we have

$$\frac{V_{filter}}{V_{buffer}} = \frac{V_4}{V_2} = \frac{V_3}{V_1} = H(f) \quad (6.1)$$

or

$$H(f) = \frac{V_{filter}}{V_{buffer}} \quad (6.2)$$

Figures 6.3 and 6.4 show simulations for the measurement technique when the filter is set to the lowest and highest bandwidths. These simulations included pad parasitics, ESD protection circuitry and a first order model for the 128 pin TQFP package used. When

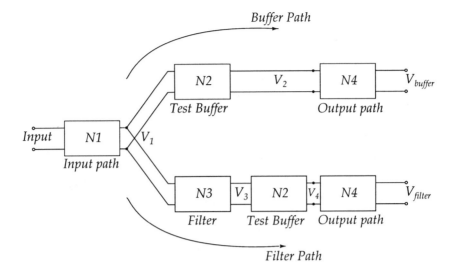

Figure 6.2. Twoport representation of the test setup.

the test buffers in the filter and buffer paths are matched, the measurement technique is exact.

Armed with (6.2), a spectrum analyzer, and a signal source, one can measure the magnitude response of the filter. To circumvent this laborious process, one can resort to a network analyzer like the HP8753D. Since the network analyzer speaks the language of scattering parameters, (6.2) will need to be interpreted in this context. Without going through parameter set conversions, substitutions and calculations that would run into a few pages of dull algebra, we simply state that

$$H(f) = \frac{S_{21,filter}}{S_{21,buffer}} \quad (6.3)$$

where $S_{21,filter}$ and $S_{21,buffer}$ are the forward scattering coefficients of the two-ports formed by the filter and buffer paths respectively.

The schematic of the test buffer is shown in Figure 6.5. It is a two stage design in order to reduce any reverse transmission. The small-sized PMOS buffers drive a differential pair biased at a very large gate overdrive voltage(for linearity). The buffers have a small input impedance with a significant capacitive component, which has been accounted for in the filter synthesis.

136   HIGH FREQUENCY CONTINUOUS TIME FILTERS

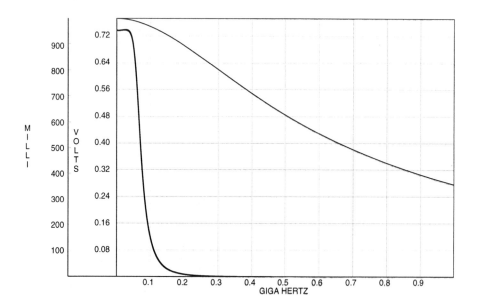

Figure 6.3. Simulation of the measurement technique when the filter bandwidth is set at the low end - the thin line is the response of the buffer path, the thick line is the actual filter response obtained by using (6.2). Since the buffer gain is almost flat in the filter passband, the response of the filter path coincides with the actual filter response and is not discernible on the plot.

2. FREQUENCY RESPONSE

All measurements were performed at room temperature. Figure 6.6 shows the pass band detail of the filter for various settings of the frequency control word. As can be expected, the response scales cleanly. In interpreting the plots, observe that the frequency axis is marked in a linear scale. Notice that the y-axis in the figure has a resolution of 1 dB per division. The relative shape of the response peaks by less than 0.2 dB **even when the filter corner is changed by 5X**. Figure 6.7 shows the stopband performance of the filter for different frequency settings. The isolation of the test setup was about 80 dB at low frequencies ($<$ 400 MHz), and steadily dropped to about 60 dB at higher frequencies. Since the test buffers and transformers have a loss of about 30 dB at high frequencies ($>$ 600 MHz), the

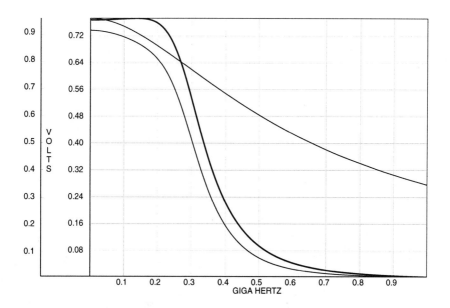

Figure 6.4. Simulation of the measurement technique when the filter bandwidth is set at the high end - the thin lines are the responses of the buffer and filter paths, the thick line is the actual filter response obtained by using (6.2).

feedthrough from the test setup makes measurements above these frequencies unreliable, so these are not shown.

The DC gain of the filter is less than unity due to the finite gain of the integrators used in the design. Figure 6.8 compares the unity-gain normalized response obtained, with an ideal fourth order Butterworth response.

Figure 6.9 shows the response of the filter, when the power supply voltage is changed from the nominal value of 3.3 V by ±10%. For this measurement, the filter bandwidth was set to 200 MHz. When the power supply voltage was varied, the bandwidth changed by ±2%. Figure 6.10 shows the response of twenty filter chips overlaid on the same plot. This is done so as to get an idea of the variation of response among different chips. For this measurement, the external frequency setting was set to a value such that the bandwidth of a particular chip was 60 MHz. Subsequent chips were measured with the same value of the external resistance. There are two reasons

138 HIGH FREQUENCY CONTINUOUS TIME FILTERS

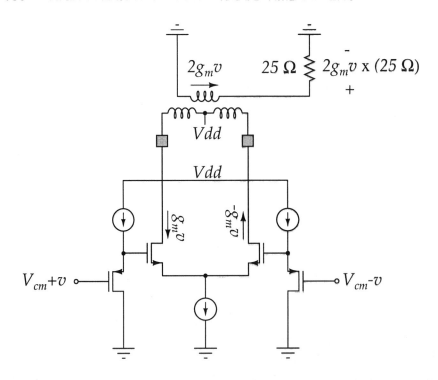

Figure 6.5. Simplified schematic of test buffer and output circuits.

for measuring the response at the low-end bandwidth. First, this corresponds to the worst case because only the minimum number (4) of unit cells are turned on. Hence mismatches in response will be most obvious. Second, at low bandwidths, the mismatch in the test buffer paths can be expected to be very small. The mean bandwidth and standard deviation of the twenty chips measured was 59.97 MHz and 0.5 MHz, respectively. These results attest to the fact that the response is very repeatable over a number of chips.

3. FILTER OUTPUT NOISE

The noise spectrum of the buffer and the filter paths was measured using a Rohde-Schwarz FSEB spectrum analyzer. In order not to get muddled with the frequency response of the test set up, we measured noise when the filter is set near the low end bandwidth. Recall from the discussions of Chapter 4 that the filter architecture is such that it has the same distortion and noise performance regardless of the filter corner frequency.

Filter Testing and Measurement Results 139

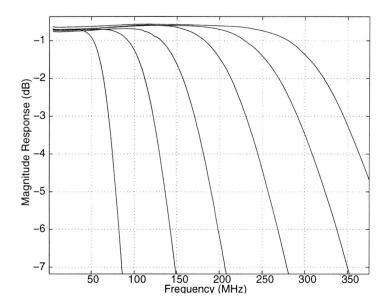

Figure 6.6. Passband detail.

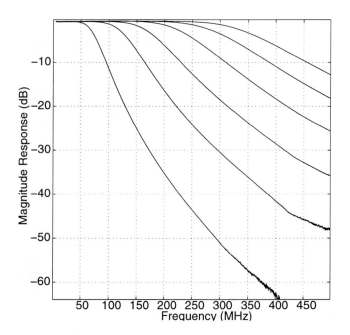

Figure 6.7. Frequency response.

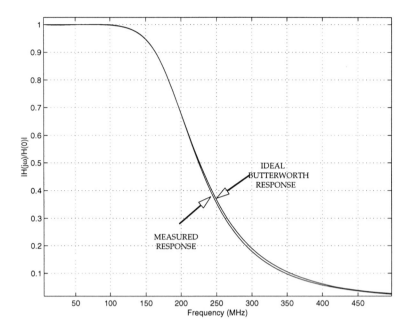

Figure 6.8. Measured response compared with an ideal 4th order Butterworth response.

We calculate the filter output noise as follows:

- All the noise from the setup is assumed to be coming from the test buffers.

- The buffer path is assumed to have a unity gain and no frequency dependence (this is approximately true within the frequency range of interest).

The noise spectra of the filter and buffer paths from 10 MHz to 150 MHz is shown in Figure 6.11. The filter output noise spectrum is shown for two different bandwidth settings - the top trace is for the case when the digital bandwidth setting code is 0000 (corresponding to a filter bandwidth of 75 MHz) while the bottom trace is for a code 0010 (bandwidth = 112 MHz). As predicted by frequency scaling theory, the output noise power spectral density of the filter at very low frequencies reduces by $10\log(112/75) \approx 1.75$ dB as the bandwidth is increased by a factor of $112/75 \approx 1.5$.

When noise powers were computed, the frequency range of 5 – 400 MHz was used. The lower limit was chosen to coincide with

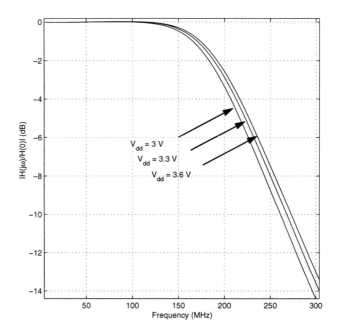

Figure 6.9. Filter response at $V_{dd} = 3\,\text{V}, 3.3\,\text{V}$ and $3.6\,\text{V}$.

the lower band of the transformers used as baluns. Notice from Figure 6.5 that if a voltage $2\Delta v$ is applied at the filter output, it will be measured as $2g_m\Delta v\,25\,\Omega$ across the spectrum analyzer input. From simulations, $g_m \approx 1/50\,\Omega$. Hence, the filter output voltage is attenuated by a factor of 2. Consider now Figure 6.1. We denote the RMS output noise of the filter by $v_{on,fil}$ and the input referred RMS noise of the testbuffer by $v_{in,buf}$. The gain from the filter output to the spectrum analyzer input is $1/2$; the integrated output noise power of the filter path was

$$\text{Integrated Noise Power}|_{filterpath} = \frac{v_{on,fil}^2 + v_{in,buf}^2}{4 \times 50} = -62.92\,\text{dBm} \tag{6.4}$$

Similarly, the output noise power of the buffer path was

$$\text{Integrated Noise Power}|_{bufferpath} = \frac{v_{in,buf}^2}{4 \times 50} = -67.42\,\text{dBm} \tag{6.5}$$

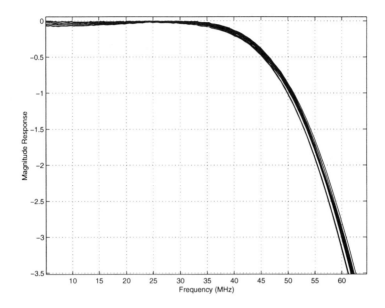

Figure 6.10. Response of 20 filters measured under identical conditions. Mean bandwidth = 59.97 MHz, Standard Deviation 0.5 MHz.

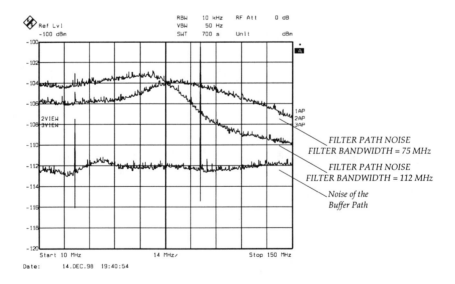

Figure 6.11. Output noise spectral density of the filter for different bandwidth settings.

Subtracting the above two equations, we get

$$v_{on,fil} \approx 257\,\mu\text{V, rms} \qquad (6.6)$$

Notice from Figure 5.36 that at 65°C, simulations predict an RMS output noise of about $275\,\mu$V. At room temperature, this corresponds to about $259\,\mu$V, rms. Thus, the measurements agree well with the value obtained from simulations.

4. DISTORTION

The total harmonic distortion of the filter as a function of peak to peak input differential voltage is shown in Figure 6.12. Measurements are for the case when the filter was set to its lowest bandwidth because random mismatches are the highest with this setting. The input tone is such that the third harmonic lies at the 3 dB bandwidth of the filter. This represents the worst case distortion condition. If a very low frequency tone is used, negligible currents flow through the integrating capacitors and the non-linearities in the filter cancel each other. If the input frequency is very high, the harmonics lie outside the filter bandwidth and hence do not manifest at the output. These observations are further supported by the distortion simulations in Figure 4.14 in Chapter 4.

5. TEMPERATURE MEASUREMENTS

Figure 6.13 shows the measured frequency response of the filter at 0°C, 45°C and 75°C. Figure 6.14 shows deviation of the filter bandwidth from the nominal as a function of temperature. The variation in the bandwidth for a 75°C change in temperature is $\pm 1.8\%$. For the same temperature range, the variation in the external resistor (to which all the transconductances in the filter are servoed) was $\pm 0.05\%$.

It was not possible to uniformly heat the complete printed circuit board - so any temperature gradients on the test board would translate to inaccuracies in the filter response. However, the path through the testbuffers and transformers had a much higher bandwidth compared to the lowest bandwidth setting of the filter. Therefore, when the filter is set to its lowest bandwidth, the mismatch in these two external paths can be expected to be small in the filter passband.

144   HIGH FREQUENCY CONTINUOUS TIME FILTERS

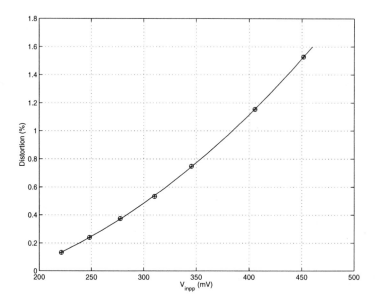

Figure 6.12.   THD as a function of input signal level ($f_{in} = f_{-3dB}/3$).

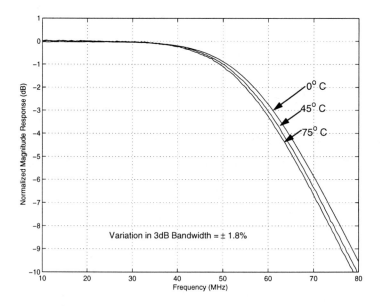

Figure 6.13.   Frequency response at $T = 0°C, 45°C$ and $75°C$.

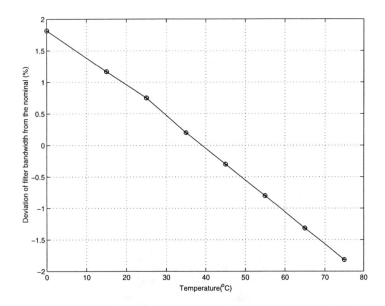

Figure 6.14. Percent deviation of filter bandwidth from the nominal value.

Hence, temperature measurements were performed only at the lowest bandwidth setting.

6. SUMMARY

Figure 6.15 shows the die photograph of the filter test chip. The active area is 0.15 mm². Since MOS capacitors are used, the integrating capacitors occupy a small fraction of the total filter area.

Table 6.1 summarizes the important measurement results. Notice that the power consumption of the filter is independent of the bandwidth setting.

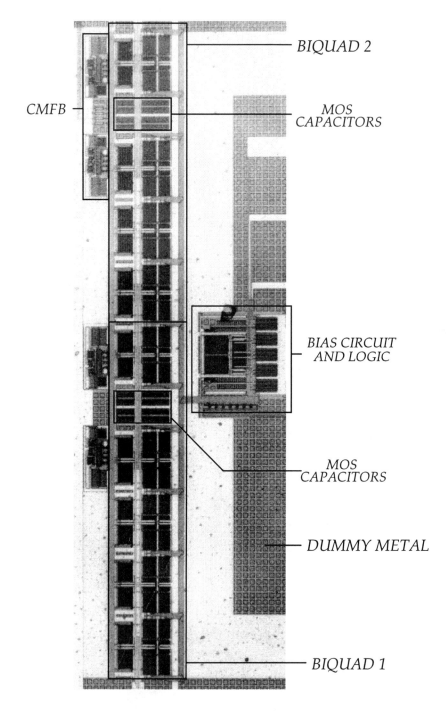

Figure 6.15. Chip Microphotograph.

Table 6.1. Summary of measured characteristics (25°C, unless noted otherwise)

Technology	0.25 μm n–well CMOS
Filter type	4th Order Butterworth
Supply voltage	3.3 V
Bandwidth programmability	60–350 MHz
Chip area	0.15 mm^2
Power[†]	70 mW
DC gain	−0.8 dB
Integrated output noise[†]	257 μV rms
f_{-3dB} variation for V_{dd} 3 − 3.6V	±2%
f_{-3dB} variation with temperature (0 − 75°C)	±1.80%
Test tone at $f_{-3dB}/3$:	
$V_{ipp,max}$ (THD ≤ −40 dB)[†]	380 mV
Dynamic range (THD ≤ −40 dB)[†]	54 dB

[†] Remains practically fixed at all bandwidths.

Chapter 7

APPLICATIONS OF SCALING TO OTHER FILTER TECHNIQUES

1. INTRODUCTION

In this chapter, we apply the constant-capacitance scaling principle to other filter architectures. First, we blend the scaling principle with the filter technique proposed by Nauta [8]. The limitations of his technique are discussed and a modified circuit architecture that overcomes its problems is proposed. Then, we discuss the application of constant-capacitance scaling to a CMOS Gm-OTA-C architecture.

2. NAUTA'S CMOS VHF FILTER TECHNIQUE

An elegant VHF filter technique based on CMOS inverters was proposed by Nauta in [8]. The philosophy of his technique was that **an absolute minimum number of signal carrying nodes must be used in order to make filters at very high frequencies possible.** This is a very sound strategy for the following reason - every node in the signal path is associated with a capacitance that has to be charged and discharged. Charging a capacitor with a voltage source with a finite output impedance takes some time - in other words, every node is associated with a pole. Since parasitic poles are not welcome, the filter should have no more than the minimum number of signal carrying nodes necessary. In a Gm-C filter, this minimum number corresponds to the number of grounded capacitors in the filter.

The simplified schematic of Nauta's scheme is shown in Figure 7.1. The CMOS inverters form the basic V-I converters. The output

150 HIGH FREQUENCY CONTINUOUS TIME FILTERS

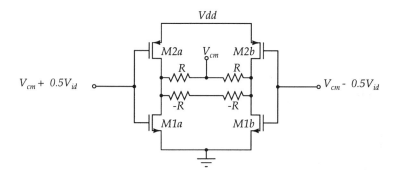

Figure 7.1. Simplified schematic of Nauta's transconductance element

common-mode voltage is maintained by using a battery of value V_{cm} along with the two resistors R. These resistors would decrease the DC gain of the transconductor. To prevent this, a negative resistance for *differential signals only*, of value $-2R$ is added in shunt with the output of the transconductor. Thus, the output nodes see a low common mode impedance and a high differential mode impedance.

The implementation of the above scheme consists of six CMOS inverters and is shown in Figure 7.2. The basic voltage to current conversion is performed by Inv1 and Inv2. Inv3 thru Inv6 control the common mode level of the output nodes *op* and *on*. Inv4 and Inv5 form the common mode voltage generators with the output resistance R. Inv3 and Inv6 form the differential negative resistances. In spite of having six inverters, notice that there are no internal signal carrying nodes in the circuit. The common-mode level of the input drive is generated using the circuit of Figure 7.3. Using the square law model for the transistors, it can be shown [8] that the transconductance of the circuit is

$$G_m = \sqrt{\mu_n \mu_p C'_{ox,n} C'_{ox,p} \left(\frac{W}{L}\right)_n \left(\frac{W}{L}\right)_p} (V_{dd} - V_{tn} + V_{tp}) \qquad (7.1)$$

where V_{tn} and V_{tp} are the threshold voltages of the *n*-channel and *p*-channel devices respectively. Tuning of the transconductance value is accomplished by changing the supply voltage Vdd.

The technique has the potential for very high frequency operation, and this has been proved by experimental results [8] - corner frequencies of the order of a hundred megahertz were achieved in a 3 μm CMOS technology. However, the technique has some drawbacks,

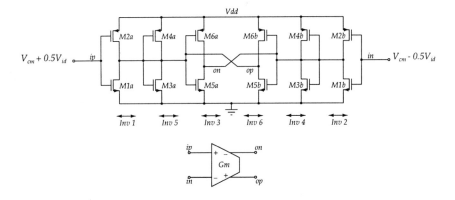

Figure 7.2. Complete schematic Nauta's transconductance element

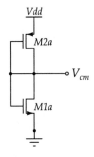

Figure 7.3. Generation of the input common-mode level

some of which are serious and make impossible its extension to the realization of programmable filters. These are described below.

- **Low DC gain :** In many modern fine-line CMOS technologies, the DC gain attainable of a single inverter is modest, especially if short channel devices are used. Moreover, the PMOS devices have a much poorer output conductance compared to the NMOS devices.

- **Transient currents from the supply voltage :** For large input signals, or if there are mismatches between the inverters, the transconductor will draw transient currents from the supply voltage. Any finite output resistance on the Vdd line could cause signals to couple from one part of the filter to another. These parasitic paths can alter the transfer function or even cause instability.

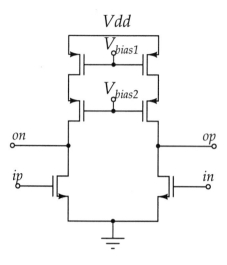

Figure 7.4. Basic transconductor principle.

- **Transconductance-dependent dynamic range :** Since the maximal signal level capable of being handled by the circuit depends on the value of transconductance realized [8], the dynamic range is not constant with tuning. The problems with this as applied to programmable filters has been discussed in detail in Chapter 3.

3. IMPROVED FILTER TECHNIQUE BASED ON NAUTA'S ARCHITECTURE

We now present a technique incorporating the constant-capacitance scaling principle along with Nauta's idea. This solves most of the problems mentioned above, while retaining the appealing qualities of Nauta's work.

3.1 BASIC TRANSCONDUCTOR

To mitigate the problem of the output conductance of the PMOS devices, an all NMOS signal path is used. Cascoded PMOS devices are used as high output impedance current sources. The basic transconductor principle is shown in Figure 7.4, where it is assumed that the input voltage has an appropriate value such that the resulting common mode output voltage keeps all devices operating in saturation. Notice that this transconductor draws a fixed current from

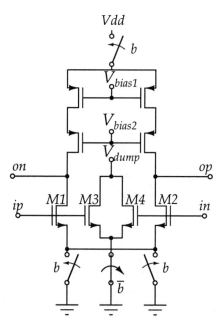

Figure 7.5. Unit transconductance element.

the power supply, avoiding cross-talk and other potential problems caused by finite impedance of the Vdd line.

3.2 TUNABILITY

To make the transconductor programmable over a wide range, the bias current can be varied. However, this is not a practical option. For instance, changing the transconductance by a factor of five would require the bias current to change by a factor of twenty five! Based on our experience with the differential pair, we use a similar strategy to generate scaled transconductors.

Consider the unit transconductance element shown in Figure 7.5. M1, M2, M3 and M4 are equal in size. M3 and M4 are dummy devices switched on when M1 and M2 are off. This ensures a constant input capacitance for the transconductor. The equivalent circuit for differential excitation when the controlling bit is zero and when it is one are shown in Figure 7.6. Notice that all nodal capacitances remain fixed while all conductances and transconductances scale by the same factor (zero in this case). There are no switches in the signal

154   HIGH FREQUENCY CONTINUOUS TIME FILTERS

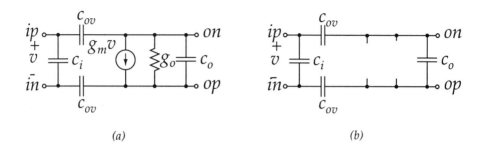

Figure 7.6. Equivalent circuits for the unit transconductance element for differential excitation when (a) $b = 0$ and (b) $b = 1$.

path. Moreover, the switches can be implemented using minimum length devices.

As discussed in Chapter 4 arrays of unit transconductors can be connected in parallel to result in a digitally tunable transconductance element with a constant excess noise factor across its tuning range. As far as noise and distortion performances go, this technique retains the same desirable features of the constant-capacitance scaled architectures presented in Chapters 4 and 5.

3.3   COMMON-MODE LEVEL SETTING

The only problem that now remains is the setting of the output common mode level. This is achieved in a manner analogous to that proposed by Nauta. The complete schematic of a unit element is shown in Figure 7.7. M1 thru M12 are equal sized devices. The input common-mode voltage is generated as shown in Figure 7.8, and is assumed to be added to the input signal.

3.4   CAPACITORS

The integrating capacitance can be made of accumulation structures. However, due to the nature of the circuit, a large portion of the capacitance is likely to be composed of the gate impedance of a saturated MOSFET. Here we estimate the distortion caused by the gate capacitance of a saturated transistor when it is driven by a sinusoidal current, as shown in Figure 7.9.

Consider the cross-section of an n-channel transistor shown in Figure 7.10 [16]. For a given set of voltages (v_G, v_S, v_D, v_B), the channel-potential as a function of the distance from the source can be

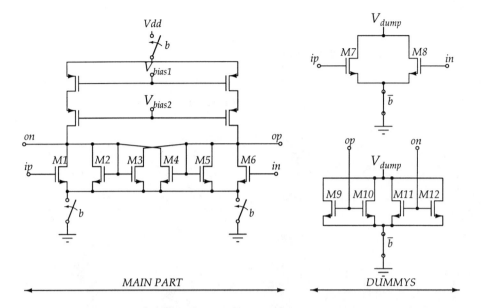

Figure 7.7. Complete unit element.

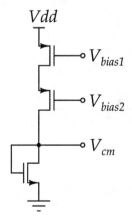

Figure 7.8. Generation of the input common-mode level.

found numerically. If the channel potential at every point along the channel is known, the total charges in the gate, bulk and inversion layer can be computed. Non-quasistatic effects were neglected and a MATLAB routine was used to numerically solve for the gate charge as a function of the gate voltage, with the body and source terminals grounded. The drain was connected to a DC potential large enough

156   HIGH FREQUENCY CONTINUOUS TIME FILTERS

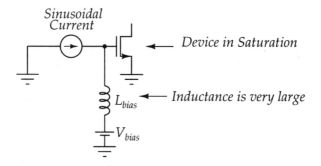

Figure 7.9.   Gate of a saturated device driven by a sinusoidal current.

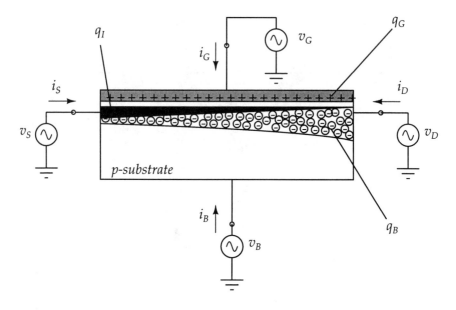

Figure 7.10. Definition of currents and charges in the presence of varying terminal voltages. Lowercase symbols with capital subscripts denote total time-varying quantities.

to ensure that the device operated in the saturation region. Under the conditions mentioned above, the charge on the gate q_G and gate capacitance C_G can be written as

$$q_G = f_G(v_G) \qquad (7.2)$$

$$C_G = \frac{dq_G}{dv_G} \qquad (7.3)$$

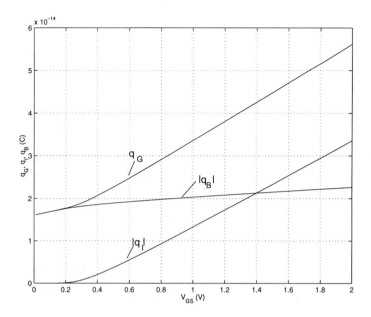

Figure 7.11. Magnitude of gate, bulk and inversion charges as a function of gate voltage ($v_D = 3\,\text{V}, v_S = v_B = 0\,\text{V}, W = 4.5\,\mu\text{m}, L = 1\,\mu\text{m}, t_{ox} = 48\,\text{Å}.$)

A plot of the magnitude of q_G, q_I and q_B as a function of v_G for a typical process is shown in Figure 7.11. The extrapolated threshold voltage for the device is about 0.3 V. C_G is shown in Figure 7.12.

The distortion in the voltage across the gate capacitance is shown in Figure 7.13. These curves are for a sinusoidal current drive, assuming that the peak voltage swing is 200 mV (corresponding to a 800 mV$_{\text{pp,diff}}$.) From the figure, the third harmonic distortion is about 0.1 percent for a gate voltage of 0.8 V.

4. A MODIFIED GM-OTA-C TECHNIQUE

In Chapter 3 we discussed the problems of the Gm-OTA-C technique when implemented in a digital CMOS technology. Here we assume that high-density poly-poly capacitors are available (for simplicity, we do not consider bottom-plate parasitics) and a parasitic insensitive integrator is desired. We show that constant-capacitance scaling can be used to obtain better performance than would otherwise be possible.

158   HIGH FREQUENCY CONTINUOUS TIME FILTERS

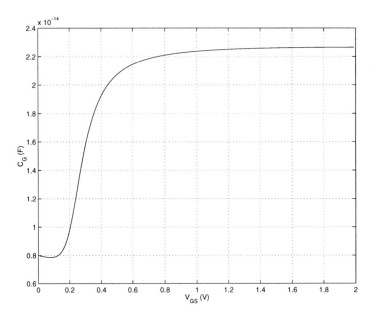

Figure 7.12.   Gate capacitance as a function of v_{GS}.

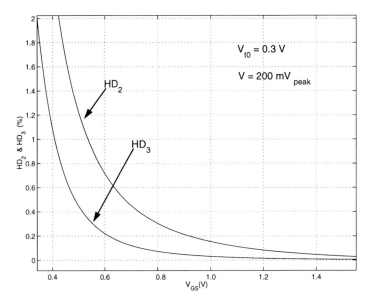

Figure 7.13.   Distortion in the gate capacitance voltage waveform.

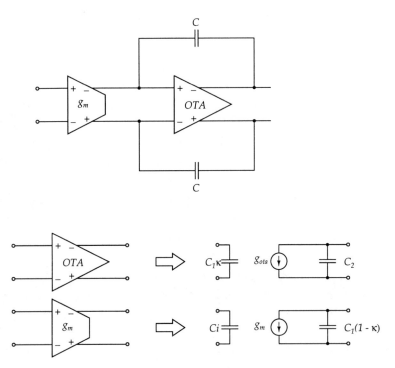

Figure 7.14. A Gm-OTA-C integrator.

A schematic of the Gm-OTA-C integrator along with simplified equivalent circuits for the transconductor and OTA are shown in Figure 7.14. Since a very high frequency of operation is desired, we assume that both are single stage designs. The difference between the terms "transconductor" and "OTA" is discussed elsewhere [22].

It can be shown that the transfer function of the integrator is

$$H(s) = \frac{\omega_i}{s}\left(\frac{1 - s/\omega_z}{1 + s/\omega_p}\right) \tag{7.4}$$

where

$$\omega_i = \frac{g_m}{C} \tag{7.5}$$

$$\omega_z = \frac{g_{OTA}}{C} \tag{7.6}$$

$$\omega_p = \frac{g_{OTA}}{(C_1 + C_2 + C_1 C_2/C)} \tag{7.7}$$

The phase error of the integrator at its unity gain frequency is given by

$$\Delta\phi = \tan^{-1}\left(\frac{\omega_i}{\omega_z}\right) + \tan^{-1}\left(\frac{\omega_i}{\omega_p}\right) \tag{7.8}$$

Note that the behavior of the phase error with ω_i follows the same pattern as discussed in Section 7 of Chapter 3. In conventional techniques, $\Delta\phi$ is minimized by increasing the ratio g_{OTA}/g_m. This pushes ω_z and ω_p far away from ω_i. To cancel the right half plane zero, a resistance of value $1/g_{OTA}$ is inserted in series with the integrating capacitance [7]. This has the disadvantage of introducing another pole in the integrator transfer function. The tracking of this resistor with the transconductance of the OTA over process and temperature is another issue.

The constant-capacitance scaling technique can be applied to the Gm-OTA-C integrator to yield a design which has constant phase error and noise performance irrespective of the unity gain frequency.

Consider the unit transconductor and unit OTA shown in Figure 7.15. The common-mode feedback circuits are not shown in order to avoid clutter in the diagram. The reader will immediately recognize the similarity to the constant-capacitance scaled Gm-C integrator of Chapter 4. When the control bit b is turned off, it can be checked that all the node capacitances remain the same, while all conductances and transconductances are scaled by the factor $\alpha = 0$. Arrays of such unit elements can be connected in parallel, to yield digitally programmable transconductors and OTAs. One such example is shown in Figure 7.16. These transconductors and OTAs can be combined to yield the integrator shown in Figure 7.17. In this figure, G_m and G_{OTA} can vary from $4g_m$ to $19g_m$ and $4g_{OTA}$ to $19g_{OTA}$, respectively. It can be shown that the phase error and noise performance of this integrator is independent of the set unity gain frequency. This can readily be inferred from our discussion on frequency scaling in Chapter 4.

5. SUMMARY AND CONCLUSIONS

In this chapter, we applied constant-capacitance scaling to Nauta's inverter-based filter design and the parasitic insensitive Gm-OTA-C architecture. While each of these techniques has its own inherent

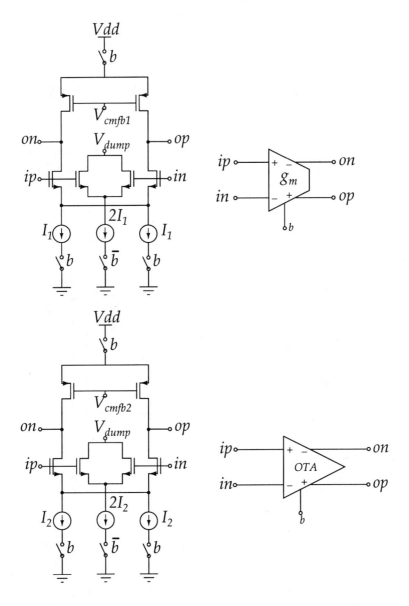

Figure 7.15. Scalable transconductor and OTA unit cells.

problems, scaling can be used to alleviate some of these. The constant-capacitance scaled versions of both these architectures, while being significant improvements, are still not as robust and efficient as the implementation described in Chapter 5. The main intention of this chapter was to show that constant-capacitance scaling can be

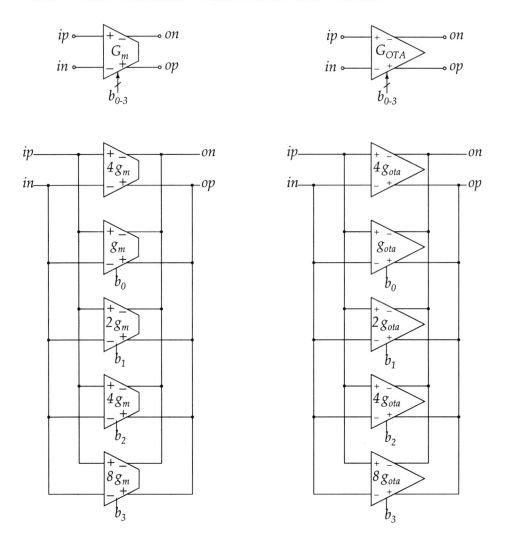

Figure 7.16. A digitally programmable transconductor and OTA.

applied to many other filter design methods, leading to improved performance.

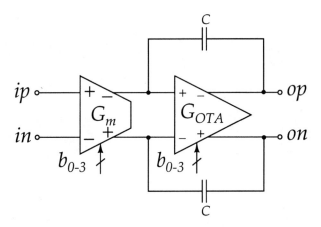

Figure 7.17. A constant-capacitance scaled Gm-OTA-C integrator.

Chapter 8

TUNING IN CONTINUOUS-TIME FILTERS

1. INTRODUCTION

The frequency response of a filter is determined by the values of transconductances, resistances and capacitances. To maintain an accurate filter response, precise *absolute* values for components are necessary. Absolute values on an integrated circuit can shift significantly from the nominal due to process parameter variations, temperature and aging. To maintain a reasonable level of insensitivity of filter characteristics due to parameter shifts, a system which corrects for these effects is required. Depending on the accuracy of response desired, many schemes with varying complexities have been proposed and implemented. All these are collectively called "automatic tuning schemes." Since tuning (even manual) of a high order filter is complex, integrated tuning schemes have generally relied on manipulating the response of basic filter building blocks like biquadratic sections. Tuning involves the following operations [43]:

- Measure the filter response (or, more precisely, some of its characteristics).
- Compare the response with the desired response.
- Apply corrective feedback to reduce the error to zero.

Depending on the complexities of the measurement and comparison techniques, an entire spectrum of tuning techniques has been

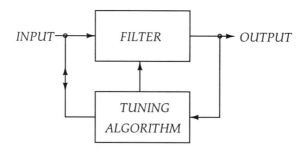

Figure 8.1. Concept of direct tuning.

developed over the years. They range from adaptive methods (very complex implementations) to "resistor-servo" techniques (simple).

The tuning strategy can be indirect [50] or direct [51]. In either case, the filter to be tuned is a voltage (or current) controlled filter, that is, a filter whose parameters are "programmable" by a set of control voltages (or currents). The concept behind direct tuning is shown in Figure 8.1. Here, the very filter that processes the signal is tuned. This can be implemented in many ways depending on the manner in which the filter is used. If continuous operation is required, the tuning algorithm should be able to infer the response of the filter from the statistics of the input and output signals. This is the idea behind adaptive tuning techniques. The tuning processor is invariably very complex. In some situations, the filter operates only in bursts. In such cases, the filter can be removed from the signal path and tuned. This kind of tuning technique has been employed successfully in video filters [52] - where tuning was performed in the field fly-back interval. Another alternative is to use two filters in a ping-pong fashion [51].

The idea behind indirect tuning is shown in Figure 8.2. The filter that processes the signal (Filter A) is left alone. Instead, the idea is to infer its parameters from measurements made on a replica filter (Filter B) resident on the same integrated circuit chip. Presumably any non-idealities in the response of the replica also exist in the main filter. If the response of the replica is corrected, the same correction when applied to the main filter should result in the desired response. In tuning parlance, the main filter is called the slave and the replica is referred to as the master. Hence, indirect tuning schemes are also referred to as master-slave tuning schemes. If a high order filter

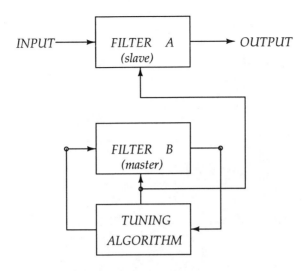

Figure 8.2. Concept of indirect tuning.

is used, using an identical replica would result in a doubling of chip area and power dissipation. An alternative scheme is to use a replica which, though simple, models the slave's behavior accurately. Many choices are possible for the master. In the simplest case, it could be a transconductor (in Gm-C filters) whose transconductance is kept constant with temperature and process variations. Here, the assumption made is that the capacitors have acceptable stability. (This was what was done in the chip described in Chapter 5.) Clearly, this results in a very simple tuning loop. Another choice is a biquadratic section, since it is the basic building block of active filters. In this case, both pole frequency (ω_0) and quality factor (Q) can be tuned. We now discuss tuning schemes where a biquad is chosen as the master. The loops which set ω_0 and Q are sometimes called Vector Lock Loops (VLL).

2. THE VOLTAGE CONTROLLED FILTER TECHNIQUE

The schematic of a second-order Gm-C filter section shown in Figure 8.3. This structure is based on the double integrator loop, and is capable of producing both low-pass and bandpass outputs. The transfer functions corresponding to these outputs are given by

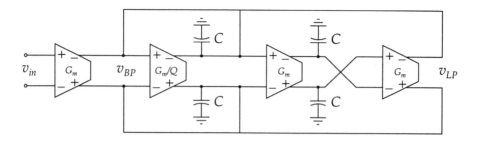

Figure 8.3. A Gm-C biquad realizing lowpass and bandpass transfer functions.

$$H_{BP}(s) = \frac{\frac{s}{\omega_0}}{\frac{s^2}{\omega_0^2} + \frac{s}{\omega_0 Q} + 1}$$

$$H_{LP}(s) = \frac{1}{\frac{s^2}{\omega_0^2} + \frac{s}{\omega_0 Q} + 1} \qquad (8.1)$$

At $s = j\omega_0$, we can write

$$H_{BP}(j\omega_0) = Q \qquad (8.2)$$
$$H_{LP}(j\omega_0) = -jQ \qquad (8.3)$$

This means that the phase difference between the lowpass (bandpass) output and the input is $90°$ ($0°$) when the filter is excited by a sinewave at its pole frequency. The filter gain under the same conditions is Q. These observations suggest the following tuning strategy:

Frequency Tuning A sinewave at the desired pole frequency is an input to the master. A negative feedback system measures the phase difference between the input and lowpass (bandpass) output and corrects ω_0 till the phase difference is $90°$ (($0°$)).

Quality Factor Tuning A negative feedback system measures the gain of the filter (at either lowpass or bandpass outputs) under the same conditions as above and changes the quality factor of the biquad to force the gain to be Q.

This kind of negative feedback system is called the Voltage Controlled Filter (VCF) loop. The general block diagram of a VLL based on a VCF is shown in Figure 8.4 [53] [54]. The variable of interest in the frequency control loop is ϕ, the phase difference

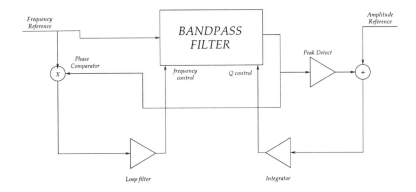

Figure 8.4. A conventional Vector Lock Loop(VLL).

between the reference and the output, while in the Q–control loop, it is M, the magnitude of the output at the pole frequency. From (8.1), we have

$$\phi(\omega_0, Q) = \frac{\pi}{2} - \arctan\left(\frac{\frac{\omega_{ref}}{Q\omega_0}}{1 - \frac{\omega_{ref}^2}{\omega_0^2}}\right) \qquad (8.4)$$

$$M(\omega_0, Q) = \frac{\frac{\omega_{ref}}{\omega_0}}{\sqrt{(1 - \frac{\omega_{ref}^2}{\omega_0^2})^2 + (\frac{\omega_{ref}^2}{\omega_0^2 Q^2})}} \qquad (8.5)$$

The above equations show the coupled nature of the phase and magnitude measurements. To make the coupling effects even more explicit, the phase and magnitude detector output surfaces are drawn in Figure 8.5. At this point we will stop by mentioning that this loop has a serious problem with coupling between the amplitude and frequency control loops. This phenomenon is discussed in a later section.

If a high frequency filter has to be tuned, we will either need an external high frequency reference or generate it on-chip using a frequency synthesizer. For this loop, it has been shown [54] that tuning accuracy is degraded if the input to the master has a high harmonic content. It is necessary, therefore to generate a clean sinewave. A square wave will not do. Also, any offset in the phase detector and the control loop will reflect as a frequency tuning error.

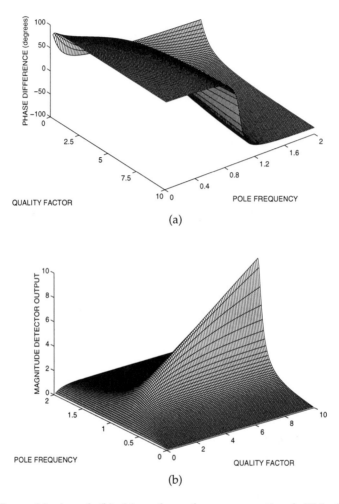

Figure 8.5. (a) ϕ and (b) M surfaces for a conventional VLL (frequency normalized to reference.)

3. THE VOLTAGE CONTROLLED OSCILLATOR TECHNIQUE

The Voltage Controlled Oscillator (VCO) technique converts a filter into an oscillator. A Phase Lock Loop (PLL) is used to lock the "filter" output to a sine wave whose frequency equals the desired pole frequency of the filter. In this method, the "master" is essentially an integrator used as the building block of the filter.

Tuning in Continuous-time Filters   171

The block schematic of the VCO loop is shown in Figure 8.6. The integrators are such that their unity gain frequency and their excess phase can be tuned by means of control voltages.

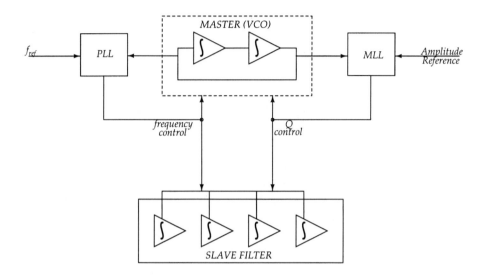

Figure 8.6.   Block diagram of a VCO tuning loop.

The frequency tuning loop uses a PLL and its functioning is well known. The Q loop works as follows : in steady state, the magnitude locked loop forces the amplitude of oscillation of the VCO to be equal to the reference amplitude. Under these circumstances, the Q-control voltage is such that each of the integrator quality factors is infinite. If this voltage is applied to the slave, each of its integrators would be ideal and thus the desired frequency response is readily obtained. However, unlike the VCF technique which corrects for the error of the whole biquad, the VCO technique (in its simplest form) only corrects for the phase error of an individual integrator in the filter. However, in the slave filter, all integrators may not be identical - hence there will be residual errors. Notice that the VCO is essentially a quadrature oscillator. If the two integrators are identical, the amplitude of oscillation at each of the VCO outputs will be identical. These two properties can be used to advantage to derive an amplitude detector which is instantaneous [55]. Care should be taken to ensure that the amplitude of oscillation is within the linear range of the integrators.

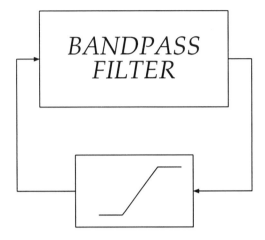

Figure 8.7. Block diagram of an oscillator.

In spite of the problem mentioned in the previous paragraph, the VCO technique has many advantages which make it more attractive than the VCF technique. First, static phase errors do not lead to frequency errors. Also, the PLL can be built such that it compares a low frequency reference to a frequency-divided version of the VCO output. The low frequency reference can be a square wave.

Later in this chapter we propose a tuning technique which has the advantages of the VCO and VCF schemes. In some sense, it is a marriage of the two. To develop the technique, we first digress to analyze a class of oscillators (this oscillator will be employed in our tuning loop) [56] [57]. The reader's patience in this matter is appreciated.

4. AN ANALYTICAL SOLUTION TO A CLASS OF OSCILLATORS

Sine-wave oscillators (Figure 8.7) contain an active element with sufficient power gain at the oscillation frequency, a frequency selective network and an amplitude stabilizing mechanism. They are capable of producing a near-sinusoidal signal with good phase noise and high spectral purity.

In a sine-wave oscillator, positive feedback is used around a frequency selective circuit to drive the poles of the corresponding closed loop linear system into the right half s-plane. In the case to

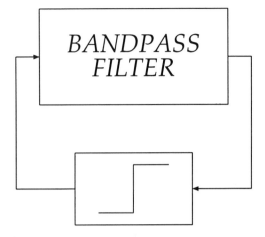

Figure 8.8. The filter comparator oscillator.

be considered in this paper, the "gain" of the amplifier is set to ∞, as shown in Figure 8.8.

Such systems are encountered in nonlinear control systems literature [58] [59] [60] and have been used by designers [61] [62] in filter tuning schemes, where one approach is to construct an oscillator with filter building blocks (integrators), for the purposes of monitoring and tuning filter characteristics. It is important in such schemes to make sure that the filter undergoes no internal limiting phenomena, so that its response can be predicted by linear system theory. This is in contrast to other oscillator methods, in which limiting within the filter can modify the frequency of oscillation [63] of the closed loop system, which then does not match and track the locations of the filter poles with variations in temperature and other environmental factors.

The system of Figure 8.8 has been studied earlier in the context of integrated oscillators using digital blocks. For an analysis of the system using non-linear differential equations, the reader is referred to [64], where the comparator is realized by using a cascade of two inverters, and the bandpass filter is a LCR series circuit. The analysis in the above work is done by approximating the non-linear transfer characteristic of the comparator by a suitable transcendental function, and solving the non-linear differential equation obtained using well known techniques.

174 HIGH FREQUENCY CONTINUOUS TIME FILTERS

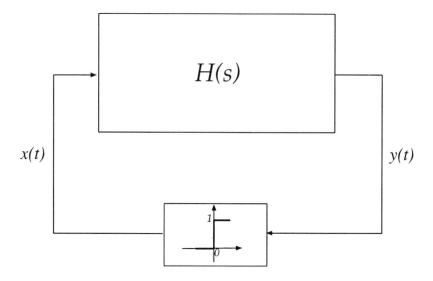

Figure 8.9. Block diagram of the oscillator.

The filter-comparator system could also be analyzed by using the describing function approach [59], where the non-linear block is replaced by an "equivalent" linear block. A first–order describing function analysis, however, predicts that the system will oscillate at the filter pole frequency, regardless of the filter quality factor, which we will see is incorrect. A higher order describing function analysis gets close to the exact result. An exact method for systems consisting of linear networks and relays has been proposed by Tsypkin and is described in detail in [59]. This method, however requires the evaluation of an infinite series using contour integrals.

The disadvantage of all the above methods for this particular system is their complexity. They do not offer much insight into system operation and the solutions are in terms of Fourier coefficients for the steady state response. In contrast, the solution we present here is straightforward, provides intuition, and gives information about all the quantities of interest (amplitude, frequency, steady state pulse shape, build–up transient) *exactly*.

The system we analyze is shown in Figure 8.9. For simplicity, until further notice we assume that the comparator output levels are 0 and 1. The filter is of the second–order bandpass type. Its transfer function is

$$H(s) = \frac{\frac{s}{\omega_0}}{\frac{s^2}{\omega_0^2} + \frac{s}{\omega_0 Q} + 1} \tag{8.6}$$

In the technique to be proposed below, we will employ the step response of the filter, $s(t)$, which is

$$s(t) = \mathcal{L}^{-1}[\frac{H(s)}{s}] = \mathcal{L}^{-1}[\frac{1}{\omega_0} \frac{1}{\frac{s^2}{\omega_0^2} + \frac{s}{\omega_0} + 1}] \tag{8.7}$$

or,

$$s(t) = \frac{1}{\sqrt{1 - \frac{1}{4Q^2}}} \exp\left(\frac{-\omega_0 t}{2Q}\right) \sin\left(\omega_0 \sqrt{1 - \frac{1}{4Q^2}}\, t\right) u(t) \tag{8.8}$$

where $u(t)$ is the unit step function. The step response crosses zero whenever

$$\sin\left(\omega_0 \sqrt{1 - \frac{1}{4Q^2}}\, t\right) = 0 \tag{8.9}$$

or at times

$$t_n = n t_1, \qquad n = 0, 1, 2... \tag{8.10}$$

where

$$t_1 = \frac{\pi}{\omega_0 \sqrt{1 - \frac{1}{4Q^2}}} \tag{8.11}$$

The mechanism of oscillation build up will be described with the aid of Figure 8.10. Let us assume that the system is initially relaxed, and that oscillation is triggered by a small positive noise at the comparator input at time $t = t_o = 0$. This will cause a step input $u(t)$ to the bandpass filter. The output $y(t)$ of the bandpass filter for $0 < t < t_1$ will coincide with the filter step response $s(t)$, as is shown in Figure 8.10. This waveform crosses zero at $t = t_1$, so at that instant the comparator switches again. Between this switching instant and the next one, the comparator output can be represented by the superposition of two steps, the first at t_0 and the second at t_1:

$$x(t) = u(t) - u(t - t_1) \tag{8.12}$$

Thus, for the same interval, the output of the linear filter can be obtained using superposition as

$$y(t) = s(t) - s(t - t_1) \tag{8.13}$$

176 HIGH FREQUENCY CONTINUOUS TIME FILTERS

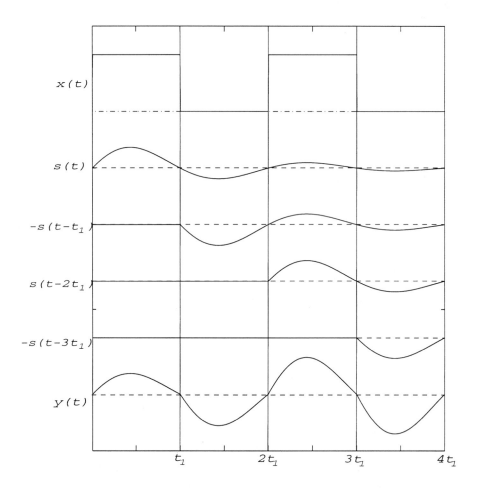

Figure 8.10. Oscillation buildup mechanism.

Notice that the zero crossings of $s(t - t_1)$ are t_1 apart from each other, just as was the case with $s(t)$. Also, the time at which $s(t - t_1)$ starts coincides with the zero crossing t_1 of $s(t)$. Thus, the output $y(t) = s(t) - s(t - t_1)$ will reach its next zero crossing when *both* $s(t)$ and $s(t - t_1)$ cross zero, i.e at $t = 2t_1$. At this point, the comparator switches again, and so until the next zero crossing, its input will be

$$x(t) = u(t) - u(t - t_1) + u(t - t_2) \tag{8.14}$$

and its output will be

$$y(t) = s(t) - s(t - t_1) + s(t - 2t_1) \tag{8.15}$$

Reasoning as above, we conclude that the next zero–crossing will occur at $t = 3t_1$, and so on. It now becomes obvious that the output of the comparator can be represented for all positive time by

$$x(t) = \sum_{n=0}^{\infty} (-1)^n u(t - nt_1) \quad , \quad t > 0 \tag{8.16}$$

The filter output then is:

$$y(t) = \sum_{n=0}^{[t/t_1]} (-1)^n s(t - nt_1) \quad , \quad t > 0 \tag{8.17}$$

where $[t/t_1]$ denotes the integer part of t/t_1. It is apparent from Figure 8.10 that the terms in the sum that produces $y(t)$ are positive for $nt_1 < t < (n+1)t_1$ if n is even, and negative if n is odd. By writing

$$\boxed{T \equiv 2t_1} \tag{8.18}$$

we see that the terms in the sum are all positive for $mT < t < mT + \frac{T}{2}$, and negative for $mT + \frac{T}{2} < t < (m+1)T$, where m is an integer. This is shown in Figure 8.11.

Steady State Response: The steady state response can be obtained by using (8.8) in (8.17) and allowing t to increase. The result of this process, as shown in Appendix E, is :

$$\begin{aligned} y_{ss}(mT + \tau) &= A \exp\left(-\frac{\omega_0 \tau}{2Q}\right) \sin(\omega_{osc}\tau), \quad 0 < \tau < \frac{T}{2} \tag{8.19} \\ &= A \exp\left(-\frac{\omega_0(\tau - \frac{T}{2})}{2Q}\right) \sin(\omega_{osc}\tau), \quad \frac{T}{2} < \tau < T \end{aligned}$$

where :

$$\omega_{osc} = \omega_0 \sqrt{1 - \frac{1}{4Q^2}} \tag{8.20}$$

178   HIGH FREQUENCY CONTINUOUS TIME FILTERS

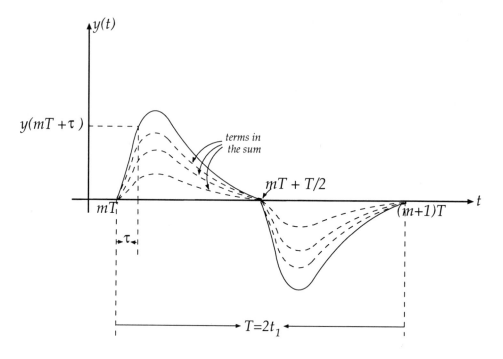

Figure 8.11.   Timing detail.

$$T = \frac{2\pi}{\omega_{osc}} \tag{8.21}$$

$$A = \frac{1}{\left[1-\exp\left(-\frac{\pi}{\sqrt{4Q^2-1}}\right)\right]\sqrt{1-\frac{1}{4Q^2}}} \tag{8.22}$$

and m is any integer. The peak of the oscillatory waveform is obtained by finding the maximum of $y_{ss}(mT + \tau)$ in the time interval $0 < \tau < \frac{T}{2}$, and is an exercise in calculus. We will find this peak for the special case when the filter quality factor is high, using (8.19) and (8.22):

$$y_{ss,peak} \cong A \cong \frac{1}{1-\exp(-\frac{\pi}{\sqrt{4Q^2-1}})} \cong \frac{1}{1-(1-\frac{\pi}{\sqrt{4Q^2-1}})} \quad , \quad Q \gg 1 \tag{8.23}$$

Thus,

$$y_{ss,peak} \cong A \cong \frac{2Q}{\pi} \quad , \quad Q \gg 1 \tag{8.24}$$

Note that, all along, we have assumed the difference in the clipping levels of the comparator to be unity. In the more general case, the output will be directly proportional to the difference in clipping levels of the comparator. This proportionality will show itself as a multiplicative constant in (8.22). If the levels of the comparator are L_1 and L_2 ($L_1 > L_2$), then the analysis just presented can still be applied, only now, the step response $s(t)$ has to be calculated with non–zero initial conditions. After going through the analysis, it can be shown that the results (8.19)–(8.21) still hold, and that the multiplicative constant that must be inserted in (8.22) is $L_1 - L_2$. This comes as no surprise, because the bandpass filter rejects the DC component of its input which is non–zero if $L_1 + L_2$ is not zero.

Startup Transient : We now consider the nature of startup dynamics assuming an initially relaxed network. For this analysis, we will focus on the peak of the oscillator output within each half period. The sequence of peaks can be considered to be a discrete time sequence, and successive peaks can be shown to occur at a time intervals of $\frac{T}{2}$ seconds. We will denote by a_n the peak of the absolute value of the output in the time interval $nt_1 < t < (n+1)t_1$. From Figure 8.10, (8.8), (8.17) and (8.18) it is evident that we can write

$$a_n = a_0 + \exp(-\frac{\omega_0 T}{4Q})a_0 + \cdots + \exp(-\frac{\omega_0 n T}{4Q})a_0$$

$$= a_0 \sum_{k=0}^{n} \exp\left(-\frac{k\omega_0 T}{4Q}\right) \quad (8.25)$$

Since

$$\sum_{k=0}^{n} x^k = \frac{x^{n+1} - 1}{x - 1} \quad (8.26)$$

we obtain

$$\boxed{a_n = a_0 \left(\frac{a^{n+1}-1}{a-1}\right)} \quad (8.27)$$

where

$$\boxed{a = \exp(-\frac{\omega_0 T}{4Q})} \quad (8.28)$$

Note that $a < 1$ and the steady state amplitude, obtained by setting $n = \infty$ in (8.27), is $\frac{a_0}{1-a}$. The time taken for the output to reach 90% of its steady state amplitude is $\frac{T}{2}(\log_a 0.1 - 1)$. This is obtained by putting $a_n = 0.9\frac{a_0}{1-a}$ in (8.27), solving for n, and calculating the required time as $nt_1 = n\frac{T}{2}$. As expected, as $a \longrightarrow 1$, (equivalent to saying that $Q \longrightarrow \infty$) the time taken to reach steady state tends to ∞.

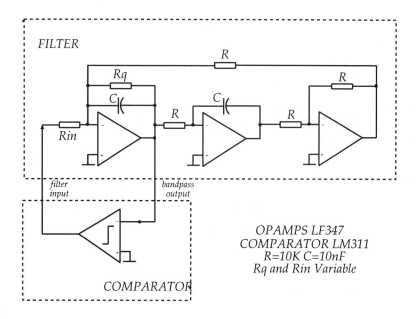

Figure 8.12. Circuit schematic.

Harmonic Distortion Analysis : Based on the detailed analysis of the oscillator steady state presented, the steady state response of the filter-comparator system can be obtained by considering an open-loop system in which the filter is driven by a periodic square wave, of frequency ω_{osc}. Expressing this square wave as a Fourier series, and calculating the attenuation offered to each Fourier component by the filter, it is straightforward [65] to calculate the spectrum at the output of the filter, and from this the Total Harmonic Distortion (THD).

Experimental Results: We now present experimental results obtained with a bread-boarded prototype of the filter-comparator system. The circuit diagram of the system is shown in Figure 8.12. The filter section is a second order op–amp RC filter, with pole frequencies in the low kHz range. The comparator used was LM311(National Semiconductor). As the quality factor was changed by varying the damping resistor of the biquad, the amplitude and frequency of oscillation changed in extremely good agreement with theoretical predictions.

In Figure 8.13, we show the predicted and observed waveforms of the comparator and the filter when Q is 1.

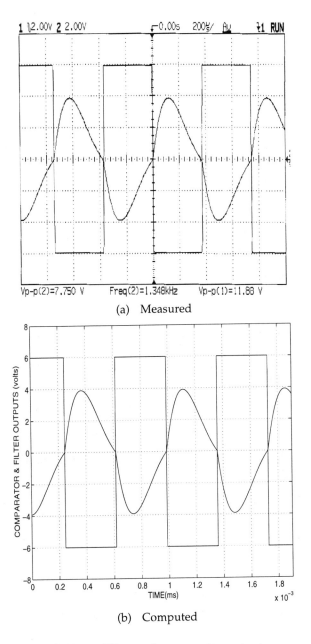

Figure 8.13. Filter and comparator outputs.

The measured and predicted amplitude of the output is shown in Figure 8.14. Note that as the quality factor increases, the amplitude increases as predicted by (8.22). The frequency and total harmonic

182   HIGH FREQUENCY CONTINUOUS TIME FILTERS

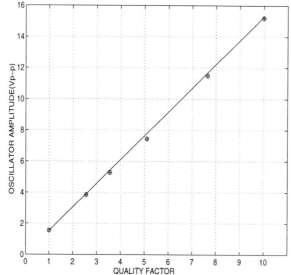

Figure 8.14. Measured($\oplus$) and predicted(—) amplitude of oscillation vs. Q.

distortion (THD) as a function of filter quality factor are shown in Figures 8.15 and 8.16 respectively. Notice that as the filter gets more selective, the harmonics of the output are attenuated to a greater degree, resulting in a lesser THD. From the figures, it is clear that experimental results agree very well with predictions.

5.   APPLICATIONS OF THE FILTER-COMPARATOR OSCILLATOR TO FILTER TUNING

In this section, we employ the filter-comparator oscillator in a Vector Lock Loop for filter tuning. The frequency and amplitude of oscillation of the oscillator are functions of the center-frequency and quality factor of the filter respectively. This suggests the following tuning strategy :

Frequency tuning A PLL is used as the frequency control loop. It servoes the frequency of oscillation of the VCO (and hence the pole frequency of the filter) to an external reference.

Q **Tuning** A negative feedback system servoes the amplitude of the bandpass output to a multiple/fraction of the square wave amplitude generated by the comparator.

Tuning in Continuous-time Filters   183

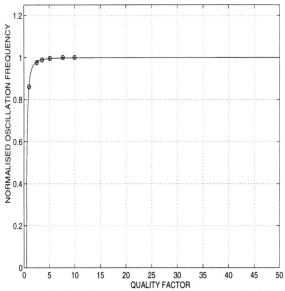

Figure 8.15. Measured($\oplus$) and predicted(—) frequency of oscillation vs. Q.

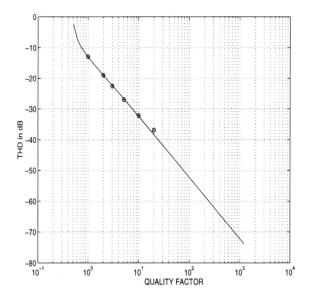

Figure 8.16. Measured($\oplus$) and predicted(—) THD vs. Q.

We will now point out the problem with inter-loop coupling in a conventional VLL which uses a second order filter. For this argument, the reader is referred to Figure 8.17. This scheme is chosen in order to appropriately introduce our proposed scheme in the sequel.

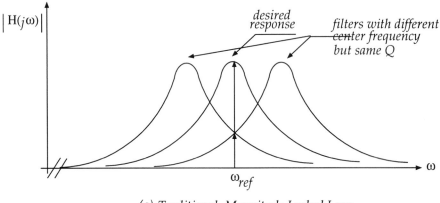

(a) Traditional Magnitude Locked Loop

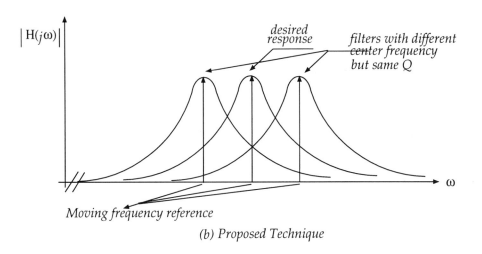

(b) Proposed Technique

Figure 8.17. Comparison of conventional and proposed technique.

Figure 8.17(a) shows the situation with the conventional vector locked loop. Assume that, to begin with, the relative shape of the response is very close to the ideal, while the center frequency deviates significantly from the desired value. For purposes of argument, assume that frequency and Q tuning is done sequentially. The magnitude detector will have an output which is very low, and this would cause the Q-loop to increase the filter Q, although there is only a frequency error in the system. Now however, when the frequency loop converges to the desired value, the quality factor will be in error, and the magnitude loop now needs more time to get back to the right value. Notice that if the desired quality factor is large, then even a small error in pole frequency could result in the magnitude detector sensing a very low output. Thus the problems with locking tend to get compounded with increasing filter selectivity. In traditional schemes, these problems are taken care of by making the Q-loop much slower than the frequency loop, so as to make the loops quasi-independent. Note that, ideally, we would want

$$\frac{\partial \phi(\omega_0, Q)}{\partial Q} = 0 \qquad (8.29)$$

$$\frac{\partial M(\omega_0, Q)}{\partial \omega_0} = 0 \qquad (8.30)$$

From Figure 8.17(a), it is obvious that all the problems with the conventional design could be avoided if we were somehow able to "move" the reference around, so that we can always sense the peak gain of the filter, no matter at what frequency it occurs. This situation is illustrated in Figure 8.17(b). Now, the magnitude detector output is constant regardless of filter center frequency, and a function of quality factor only. To generate a "reference frequency" which is always equal to the filter pole frequency, one can excite the filter and pass its output through a limiter to obtain a constant amplitude. This is precisely what the system of Figure 8.9 does. From (8.22) it is apparent that the amplitude of oscillation is now a single valued function of filter quality factor only, and is completely independent of pole frequency.

The entire vector lock loop is shown in Figure 8.18. The pole frequency of the filter is set by locking the oscillation frequency to the reference using a phase–lock loop. The quality factor is set by measuring output magnitude. From (8.20), we see that the oscillator frequency is an extremely weak function of Q. As a numerical example, the difference between the oscillation frequency when Q changes from 5 to 20 (a change in Q of 300 %) is just 2%. Thus, we can conclude that the oscillation frequency is essentially independent of Q for reasonably high values of Q. The frequency and amplitude of oscillation as a function of normalized pole frequency and quality factor are shown in Figure 8.19.

186   HIGH FREQUENCY CONTINUOUS TIME FILTERS

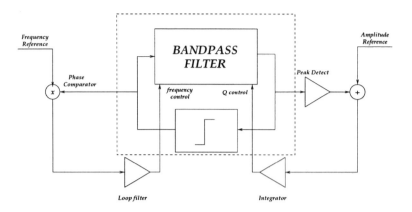

Figure 8.18.   Proposed VLL.

The independence of magnitude and frequency measurements is apparent from these two surface plots. We now discuss how the scheme just presented is different from classical VCO methods discussed in the literature. The classical methods also use the PLL principle and amplitude stabilization, but they focus on single integrators, as opposed to the biquadratic section in the proposed scheme. The VCOs implemented in the traditional schemes limit amplitude of oscillation using non–linear methods, but assume that the frequency of oscillation remains that of the resonator, which is incorrect. Although the oscillation frequency will be close to the frequency of the resonator, it will nevertheless be dependent on the nature of the non–linearity of the amplitude stabilizing element [63] [61]. In indirect tuning methods, this makes tight tracking between master and slave difficult. Our scheme operates the filter within its limits of linearity, and can be used around a resonator with any Q greater than 0.5. An alternate solution to keep the filter operating in a linear mode is to use an AGC circuit instead of the comparator in Figure 8.9. Tuning of infinite Q filters by this method has been proposed in [55], [66] and [67]. Note that even in this case, the amplitude and frequency loops are independent [55]. This is more complicated to implement than the comparator method, and does not offer us the convenience of a square wave output (which is readily available in the comparator case). We now summarize the advantages of the VLL just presented.

- The pole frequency can be tuned with absolutely no error in spite of offsets in the frequency control loop because the system utilizes the PLL principle, in which phase errors do not result in frequency errors.

- The reference can be a square wave, unlike in the VCF case, which demands a reference signal with low harmonic content.

- The filter operates in a linear fashion, and the oscillation frequency of the entire system tracks the pole frequency of the filter in the presence of variations in ambient conditions and other environmental factors.

- The amplitude and frequency loops are independent.

- The scheme can be used in direct tuning schemes, because the filter to be used can be tuned directly in contrast to conventional VCO schemes which tune individual integrators.

Thus, this loop is a marriage of the VCF and the PLL schemes, combining the advantages of both in the same method, and getting rid of the disadvantages of either methods. The loop has the same circuit complexity as any other VLL scheme.

Experimental Results: A low-frequency version of the proposed Vector Locked Loop was bread-boarded. The Master–Slave system was realized by using MOS transistor arrays. The filter topology was a second order filter of the Tow-Thomas kind, with tunable pole frequency and quality factor. The comparator used was an LM311 (National Semiconductor Corp.). The pole frequency and quality factor of the filter were observed to adjust to the reference frequency and the DC voltage reference to the Magnitude Locked Loop. Setting these quantities, filter tuning could be accomplished for reference frequencies of 1.4 kHz to 2.7 kHz, and Q values from 1 to 6. No special steps were adopted for stabilizing the loops, and we encountered no problems with stability of either loop. The limited capture and lock ranges of the PLL were due to the fact that no attempt was made to optimize the design. Figure 8.20 shows the functionality of the frequency and Q loops.

6. CONCLUSION

In this chapter, we have discussed various tuning strategies for integrated continuous-time filters. We presented an analytical technique for the solution of a class of sinusoidal oscillators. A vector lock loop, based on this class, has been proposed. The individual loops of this VLL are uncoupled. This scheme combines the best of the VCF and VCO schemes.

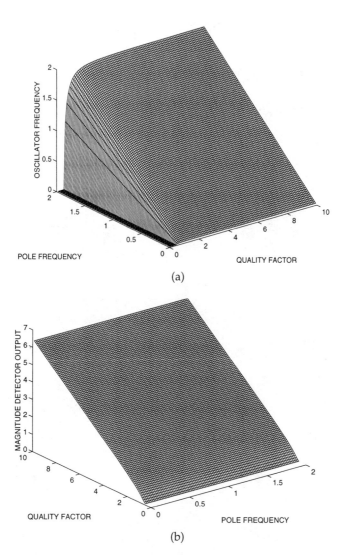

Figure 8.19. (a) Frequency and (b) Amplitude surfaces for proposed VLL (frequency normalized to desired pole frequency.)

190   HIGH FREQUENCY CONTINUOUS TIME FILTERS

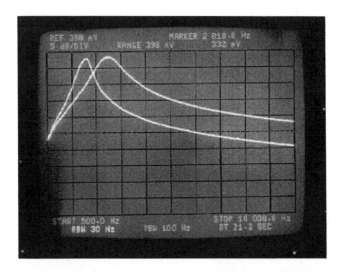

(a) Varying center frequency by varying frequency reference.

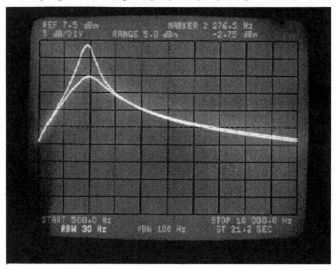

(b) Varying quality factor by varying voltage reference.

Figure 8.20.   Functionality testing of the proposed VLL.

References

[1] A. A. Abidi, "Integrated circuits in magnetic disk drives," *Digest of the European Solid State Circuits Conference*, Ulm, pp. 48-57, Sept. 1994.

[2] K. W. Moulding and G. A. Wilson, "A fully integrated five-gyrator filter at video frequencies," *IEEE Journal of Solid-State Circuits*, Vol. 13, pp. 303-7, June 1978.

[3] K. S. Tan and P. R. Gray, "Fully integrated analog filters using bipolar-JFET technology," *IEEE Journal of Solid-State Circuits*, Vol. 2, pp. 814-21, Dec. 1978.

[4] Y. P. Tsividis and J. O. Voorman (editors), "Integrated Continuous-time Filters - Principles, Design and Applications," *IEEE Press*, Piscataway, 1992.

[5] J. M. Khoury, "Design of a 15-MHz CMOS continuous-time filter with on-chip tuning," *IEEE Journal of Solid-State Circuits*, Vol. 26, pp. 1988-1997, Dec. 1991.

[6] G. A. DeVeirman and R. G. Yamasaki, "Design of a bipolar 10-MHz programmable continuous-time equiripple linear phase filter," *IEEE Journal of Solid-State Circuits*, Vol. 27, pp. 324-331, Mar. 1992.

[7] C. R. Laber and P. R. Gray, "A 20-MHz sixth order parasitic-insensitive continuous-time filter and second-order equalizer optimized for disk-drive read channels," *IEEE Journal of Solid-State Circuits*, Vol. 28, pp. 462-470, Apr. 1993.

[8] B. Nauta, "A CMOS transconductance-C filter technique for very high frequencies," *IEEE Journal of Solid-State Circuits*, Vol. 27, pp. 142-153, Feb. 1992.

[9] I. Mehr, D. Welland, "A CMOS continuous-time Gm-C filter for PRML read channel applications at 150 Mb/s and beyond,"*IEEE Journal of Solid State Circuits*, Vol. 32, No. 4, pp.499-513, April 1997.

[10] F. Rezzi, I. Bietti, M. Cazzaniga and R. Castello, "A 70-mW seventh-order filter with 7-50 MHz cutoff frequency and programmable boost and group delay equalization ," *IEEE Journal of Solid-State Circuits*, Vol. 32, pp. 1987-1999, Dec. 1997.

[11] Y. Wang and G. Uehara, "A 3-V high-bandwidth integrator for magnetic disk read channel continuous-time filtering applications," *Digest of Technical Papers*, CICC, May 1998.

[12] V. Gopinathan, M. Tarsia and D. Choi, "A 2.5 Volt, 30-100 MHz 7th order equiripple group delay filter in 0.25μ CMOS technology," *Digest of Technical Papers*, ISSCC, San Francisco, February 1999.

[13] Q. Huang,"A MOSFET-only continuous-time bandpass filter,"*IEEE Journal of Solid State Circuits*, Vol. 32, No.2, February 1997.

[14] A. Behr, M. Schneider, S. Filho and C. Montoro, "Harmonic distortion caused by capacitors implemented with MOSFET gates,"*IEEE Journal of Solid State Circuits*, Vol. 27, No.10, October 1992.

[15] S. Pavan, Y. Tsividis and K. Nagaraj,"Modeling of Accumulation MOS Capacitors for Analog Design in Digital VLSI Processes," *IEEE International Symposium on Circuits and Systems*, Vol. 1, pp. 249-252, Jun 1-3 1999, Orlando, Florida.

[16] Y. P. Tsividis, "Operation and Modeling of the MOS Transistor,"*McGraw Hill Book Company*, New York, New York; Second edition, 1999.

[17] N. D. Arora, R. Rios and C.-L. Huang, "Modeling the polysilicon-gate depletion in MOS structures,"*IEEE Electron Device Letters*, Vol. ED-39, pp. 932-938, 1994.

[18] B. Ricco, R. Versari and D. Esseni, "Characterization of polysilicon depletion effect and its impact on submicrometer CMOS circuit performance,"*IEEE Transactions on Electron Devices*, Vol. 42, pp.935-943, May 1995.

[19] R. Rios and N. D. Arora, "Determination of ultra-thin gate oxide thickness for CMOS structures using quantum effects,"*IEDM Technical Digest*, pp.613-616, 1994.

[20] F. Stern and W. E. Howard, "Properties of semiconductor surface inversion layers in the electric quantum limit,"*Physical Revue B*, Vol. 163, pp.816-835, 1967.

[21] M. Ghausi and J. Kelly, "Introduction to Distributed-Parameter Networks," *Holt, Rinehart and Winston, Inc.*, 1973.

[22] Y. Tsividis, "Integrated continuous-time filter design – an overview,"*IEEE Journal of Solid State Circuits*, Vol. 29, No. 3, pp. 166-176, March 1994.

[23] T. Georgantas, Y. Papananos and Y. P. Tsividis, " A comparitive study of five integrator structures for monolithic continuous-time filters,"*Proceedings of the IEEE International Symposium on Circuits and Systems*, pp. 1259-1262, June 1993.

[24] M. Banu and Y. P. Tsividis, "Fully integrated active RC filters in MOS technology," *IEEE Journal of Solid-State Circuits*, Vol. 18, pp. 644-51, Dec. 1983.

[25] M. Banu and Y. P. Tsividis, "An elliptic continuous-time CMOS filter with on-chip automatic tuning," *IEEE Journal of Solid-State Circuits*, Vol. 20, pp. 1114-21, Dec. 1985.

[26] M. Banu and Y. P. Tsividis, "Detailed analysis of nonidealities in MOS fully integrated active RC filters based on balanced networks," *IEE Proceedings, Part G*, Vol. 131, pp. 190-6, Oct. 1984.

[27] Y. P. Tsividis, M. Banu and J. M. Khoury, "Continuous-time MOSFET-C filters in VLSI," *IEEE Journal of Solid-State Circuits*, Vol. 21, pp. 15-30, Feb. 1986.

[28] H. Khorramabadi, P. R. Gray, "High frequency CMOS continuous-time filters," *IEEE Journal of Solid-State Circuits*, Vol. 19, pp. 939-48, Dec. 1984.

[29] S. Willingham, K. Martin and A. Ganesan, "A BiCMOS low-distortion 8-MHz low-pass filter ," *IEEE Journal of Solid-State Circuits*, Vol. 28, pp. 1234-45, Dec. 1993.

[30] F. Krummenacher and N. Joehl, "A 4-MHz CMOS continuous-time filter with on-chip automatic tuning," *IEEE Journal of Solid-State Circuits*, Vol. 23, pp. 750-8, June. 1988.

[31] Z. Czarnul, Y. Tsividis and S. C. Fang, "MOS transconductors and integrators with high linearity," *Electronics Letters*, Vol. 22, pp. 245-6, 1986.

[32] J. Pennock, P. Frith and R. Barker, "CMOS Triode transconductor continuous time filters," *Proceedings of IEEE Custom Integrated Circuits Conference*, pp. 378-81, May. 1986.

[33] R. Alini, A. Bashirotto and R. Castello, "8-32 MHz tunable BiCMOS continuous time filter," *Proceedings of Euorpean Solid State Circuits Conference*, pp. 9-12, 1991.

[34] G. Groenewold, "Optimal dynamic range integrators," *IEEE Transactions on Circuits and Systems - Part I*, Vol. 39, pp. 614-27, Aug. 1992.

[35] S. Pavan, Y. Tsividis and K. Nagaraj, "Widely programmable high frequency continuous-time filters in digital CMOS technology," *IEEE Journal of Solid-State Circuits*, to appear in April 2000.

[36] S. Pavan and Y. Tsividis, "Time-scaled electrical networks - properties and applications in the design of programmable analog filters," *IEEE Transcations on Circuits and Systems - Part II* , to appear in 2000.

[37] A. Sedra and P. Brackett, "Filter Theory and Design : Active and Passive,"*Matrix Publishers*, Beaverton, Oregon.

[38] L. Toth, G. Efthivoulidis, V. Gopinathan and Y. P. Tsividis, "General results for resistive noise in active RC and MOSFET-C filters," *IEEE Transcations on Circuits and Systems - Part II* , vol. 42, pp. 785-93, Dec. 1995.

[39] L. Chua, C. Desoer and E. Kuh, "Linear and Nonlinear Circuits,"*McGraw Hill Publishing Company*, New York, 1987.

[40] A. M. Durham, W. Redman-White and J. B. Hughes, "Low distortion VLSI compatible self-tuned continuous-time monolithic filters," *Proceedings of IEEE International Symposium on Circuits and Systems* , pp. 1448-51, May. 1991.

[41] H. Khorramabadi, M. Tarsia and N. S. Woo, "Baseband filters for IS-95 CDMA receiver applications featuring digital automatic frequency tuning," *Digest of Technical papers, ISSCC*, pp. 142-3, Jan. 1996

[42] S. Pavan, Y. Tsividis and K. Nagaraj, " A 60-350 MHz programmable analog filter in a digital CMOS process," *Proceedings of the European Solid State Circuits Conference*, pp. 46-49, Sept. 1999.

[43] R. Schaumann, K. Laker, M. Ghausi, " Analog Filter Design - Passive, Active RC and Switched-Capacitor,"*Prentice Hall Publishers*, Englewood Cliffs, 1992.

[44] K. R. Lakshmikumar, R. Hadaway and M. A. Copeland , " Characterisation and modeling of mismatch in MOS transistors for precision analog design," *IEEE Journal of Solid-State Circuits*, Vol. 21, pp. 1057-66, Dec. 1986.

[45] M. J. Pelgrom, A. L. Duinmaijer and C. J. Welbers, "Matching properties of MOS transistors,"*IEEE Journal of Solid-State Circuits*, Vol. 24, pp. 1433-40, Oct. 1989.

[46] H. Tuinhout, M. J. Pelgrom and M. Vertregt, "Effects of metal coverage on MOSFET matching,"*Proceedings of the Internation Electron Devices Meeting*, pp. 735-38, Dec. 1996.

[47] Y. P. Tsividis, "Mixed Analog-Digital VLSI Devices and Technology,"*McGraw Hill Book Company*, New York, 1996.

[48] R. Zele, D. Allstot, "Low power CMOS continuous-time filters," *IEEE Journal of Solid State Circuits*, Vol. 31, No. 2, Feb. 1996, pp. 157-168.

[49] K. Sakurai and N. Tamaru, "Simplified interconnect formulas", *IEEE Transactions on Electron Devices*, Vol. 30, No. 2, February 1983, pp. 183-188.

[50] K. R. Rao, V. Sethuraman and P. K. Neelakantan, "Novel follow-the-master filter,"*Proceedings of the IEEE*, Vol. 63 , pp.1725-1726 ,Dec. 1977.

[51] Y. Tsividis, "Self–tuned filters,"*Electronics Letters* , vol. 17, no. 12, pp. 406–407, June. 1981.

[52] R. Koblitz and M. Rieger, "A BiCMOS TV-Signal processor ," in *Integrated Continuous-Time Filters: Principles, Design and Applications*, edited by Y. Tsividis and J. Voorman, IEEE Press, New York, 1992.

[53] D. Senderowicz, D. A. Hodges and P. R. Gray, "An NMOS integrated vector–locked loop,"*Proceedings of IEEE International Symposium on Circuits and Systems*, 1982, pp. 1164-1167.

[54] V. Gopinathan, Y. Tsividis, K–S. Tan and R. K. Hester, "Design considerations for high-frequency continuous-time filters and implementation of an anti-aliasing filter for digital video,"*IEEE Journal of Solid State Circuits*, Vol. SC-25, no. 6, pp. 1368-1378, Dec. 1990.

[55] J. O. Voorman, "Balanced gyrator filters,"*Integrated Continuous Time Filters–Design and Applications*, IEEE Press, p.83, 1992.

[56] S. Pavan and Y. Tsividis, "An analytical solution to a class of oscillators and its application to filter tuning," *IEEE Transactions on Circuits and Systems - Fundamental Theory and Applications* , Vol. 47, no. 5, pp. 547-556, May 1998.

[57] S. Pavan and Y. Tsividis, "An analytical solution to a class of oscillators and its application to filter tuning," *Proceedings of the IEEE International Symposium on Circuits and Systems*, Vol. 1, pp. 249-252, Jun 1-3 1998, Monterey, California.

[58] T. Stern, "Theory of Nonlinear Networks and Systems,"*Addison–Wesley Publishing Company*, Reading, Mass. 1965.

[59] A. Gelb and Vander Velde, "Multiple-Input Describing Functions and Nonlinear System Design,"*McGraw–Hill Publishing Company* , New York 1968.

[60] Ya. Tsypkin, "Relay Automatic Systems," *Nauka* , Moscow, 1974.

[61] J. Khoury, "Design of a 15 MHz CMOS continuous-time filter with on-chip tuning,"*IEEE Journal of Solid State Circuits*, Vol. SC-26, no. 12, pp. 1988-1997, Dec. 1991.

[62] D. Welland, S. Phillip, Ka Y. Leung, G. Tuttle, S. Dupuie, D. Holberg, R. Jack, N. Sooch, K. Anderson, A. Armstrong, R. Behrens, W. Bliss, T. Dudley, W. Foland, N. Glover and L. King, "A digital read-write channel with EEPR4 detection,"*IEEE International Solid State Circuits Conference*, Digest of Technical Papers pp. 276-277, 1994.

[63] R. Adams and D. Pederson, "Nonlinear contribution to oscillation-frequency sensitivity in RC integrated oscillators,"*IEEE Journal of Solid State Circuits*, Vol. SC-6, no. 12, pp. 406-412, Dec. 1971.

[64] M. Murata, M. Ohta, K. Suzuki and T. Namekawa, "Analysis of an oscillator consisting of digital circuits,"*IEEE Journal of Solid State Circuits*, Vol. SC-5, no. 8, pp. 165-168, Aug. 1970.

[65] K. Clarke and D. Hess, "Communication Circuits: Analysis and Design",*Addison–Wesley Publishing Company*, Reading, Mass. 1971.

[66] V. Gopinathan, "High frequency transconductance–capacitance continuous–time filters," Ph.D Dissertation, Columbia University, 1990.

[67] J. Khoury, Private Communication.

Appendix A
AN EXPLICIT EXPRESSION FOR ψ_s IN A MOS ACCUMULATION CAPACITOR

In order to find an explicit expression for ψ_s in terms of V_{GB}, we rewrite (2.23) as

$$\frac{V_{GB} - V_{FB}}{\phi_t} = \frac{\psi_s}{\phi_t} + \frac{\gamma}{\sqrt{\phi_t}}\sqrt{\exp\left(\frac{\psi_s}{\phi_t}\right) - \frac{\psi_s}{\phi_t} - 1} \qquad (A.1)$$

Working with normalized variables

$$z = \frac{V_{GB} - V_{FB}}{\phi_t} \qquad (A.2)$$

$$x = \frac{\psi_s}{\phi_t} \qquad (A.3)$$

$$a = \frac{\gamma}{\sqrt{\phi_t}} \qquad (A.4)$$

we have

$$z = x + a\sqrt{e^x - x - 1} \qquad (A.5)$$

We need to find x in terms of z. Notice that the only parameter in the equation is a. If we could find an approximate explicit equation for x for practical values of a, the surface potential will "track" changes in substrate doping concentration and temperature. Then, the model formulation is based entirely on the physics of the device, and is devoid of function fitting for particular values of process parameters and temperature.

As a guide towards an approximation, we consider the asymptotic behavior of (A.5). For small x, the exponential can be approximated by the first three terms of its Taylor expansion, whereas for large x it becomes dominant. Thus, we obtain

$$z \approx \begin{cases} x + \frac{ax}{\sqrt{2}}, & \text{small } x \\ a \exp(x/2), & \text{large } x \end{cases} \quad (A.6)$$

Inverting these relations, we obtain:

$$x \approx \begin{cases} \left(\frac{a\sqrt{2}}{a+\sqrt{2}}\right) \frac{z}{a}, & \text{small } z \\ 2 \log(z/a), & \text{large } z \end{cases} \quad (A.7)$$

The second factor in these relations can be approximated by $\log(1 + z/a)$, for both small and large z. The first factor is typically 1 for small z, and becomes 2 for large z. This behavior can be satisfied by the function $2 \left(\frac{z+k_1}{z+k_2}\right)$, with k_1 and k_2 typically being 3 and 6, respectively. Thus the behavior in (A.7) can be approximated by:

$$x \approx 2 \left(\frac{z+k_1}{z+k_2}\right) \log\left(1 + \frac{z}{a}\right) \quad (A.8)$$

Although this relation was developed from the asymptotic behavior of (A.5), we find that it approximates the latter well for the entire useful range of z and for typical values of a, when $k_1 = 3$ and $k_2 = 6$ are chosen. Thus, rewriting (A.8) by using (A.2–A.4), we obtain (2.25).

Appendix B
CALCULATION OF THE BIAS VOLTAGE AT WHICH $\frac{dC_{GB}}{dV_{GB}} = 0$

In this appendix we calculate an approximate value for V_{GB} at which the C–V curve has a "flat-top" ($\frac{dC_{GB}}{dV_{GB}} = 0$). Let this occur for a $V_{GB} = V_{GB0}$. Differentiating (2.48), we get

$$\left.\frac{d(1/C'_{dep})}{dV_{GB}}\right|_{V_{GB0}} = -\left.\frac{d(1/C'_c)}{dV_{GB}}\right|_{V_{GB0}} \quad (B.1)$$

For typical process parameters, when $V_{GB} = V_{GB0}$, $V_{GB0} \gg \psi_s \gg \phi_t$. Hence, (2.29) and (2.25) can be approximated as

$$C'_c = \frac{\gamma C'_{ox}}{2\sqrt{\phi_t}} \exp(\frac{\psi_s}{2\phi_t}) \quad (B.2)$$

and

$$\psi_s = 2\phi_t \log\left(\frac{V_{GB} - V_{FB}}{\gamma\sqrt{\phi_t}}\right) \quad (B.3)$$

Using (B.3) in (B.2), we get

$$C'_c = \frac{C'_{ox}}{2\phi_t}(V_{GB} - V_{FB}) \quad (B.4)$$

Using (2.50),

$$\left.\frac{d(1/C'_{dep})}{dV_{GB}}\right|_{V_{GB0}} = \frac{2}{\gamma_p^2 C'_{ox}}\left(1 - \frac{d\psi_s}{dV_{GB}}\right) \quad (B.5)$$

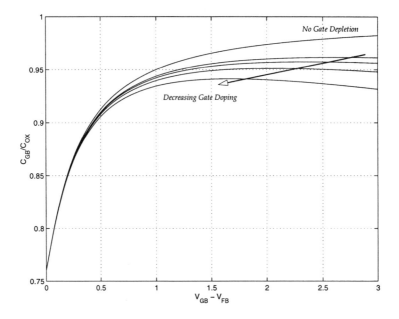

Figure B.1. Simulation of the effect of finite polysilicon doping on C–V characteristics.

From (B.3),

$$\frac{d\psi_s}{dV_{GB}} = \frac{2\phi_t}{V_{GB} - V_{FB}} \tag{B.6}$$

This leads to

$$\left.\frac{d(1/C'_{dep})}{dV_{GB}}\right|_{V_{GB0}} = \frac{2}{\gamma_p^2 C'_{ox}}\left(1 - \frac{2\phi_t}{V_{GB} - V_{FB}}\right) \tag{B.7}$$

From (B.4),

$$\left.\frac{d(1/C'_c)}{dV_{GB}}\right|_{V_{GB0}} = -\frac{2\phi_t}{C'_{ox}(V_{GB0} - V_{FB})^2} \tag{B.8}$$

Assuming that $\frac{2\phi_t}{V_{GB} - V_{FB}} \ll 1$ (we are only interested in *estimating* V_{GB0}) and using (B.7) and (B.8) in (B.1), we get

$$\frac{2}{\gamma_p^2 C'_{ox}} = \frac{2\phi_t}{C'_{ox}(V_{GB0} - V_{FB})^2} \tag{B.9}$$

or

$$V_{GB0} \approx V_{FB} + \gamma_p \sqrt{\phi_t} \tag{B.10}$$

Appendix B: CALCULATION OF THE BIAS VOLTAGE AT WHICH $\frac{dC_{GB}}{dV_{GB}} = 0$

The above relation is within a few hundred millivolts of an accurate value obtained from simulations. Figure B.1 shows the C-V curve of a MOS capacitor in accumulation, as its gate doping is progressively reduced. Notice that the "flat-top" occurs at lower bias voltages as the gate doping concentration is reduced.

Appendix C
SUMMARY OF THE PROPERTIES OF A DISTRIBUTED RC LINE

We summarize the properties of a one-dimensional uniformly distributed RY ($\overline{URY}$) structure. The meaning of $\overline{URY}$ is illustrated

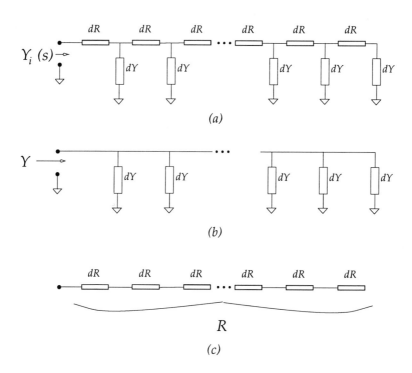

Figure C.1. A uniformly distributed RY line.

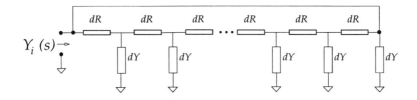

Figure C.2. Admittance of a $\overline{URY}$ line contacted at both ends.

in Figure C.1. The line has a total resistance R and total admittance Y, distributed as shown in Figure C.1(a). Notice that $Y(s)$ is the value of $Y_i(s)$ when R is zero (Figure C.1(b)). R is the total resistance of the line (Figure C.1(c)). In Figure C.2, the $\overline{URY}$ network is contacted at both ends. This is relevant to the way contacts are made to the capacitor plates, as in Figure 2.15a. For this connection, it can be shown that the driving point admittance $Y_i(s)$ is [21]

$$Y_i(s) = \frac{4}{R}\sqrt{\frac{RY(s)}{4}} \tanh\left(\sqrt{\frac{RY(s)}{4}}\right) \qquad (C.1)$$

Appendix D
SMALL SIGNAL MOS TRANSISTOR MODELS

In this appendix we review the small-signal equivalent circuit for the MOS transistor [16]. First, we consider the device part between the source and drain, containing the inversion layer, the depletion region, the oxide and the gate. This part is shown as the part within broken lines in Figure D.1 and is called the intrinsic part of a transistor. The rest of the device is called the extrinsic part and is responsible for parasitic effects.

1. OPERATION IN STRONG INVERSION AND SATURATION

The DC model for the device in is given to first order by

$$I_{DS} = \frac{\mu C'_{ox}}{2a} \left(\frac{W}{L}\right)(V_{GS} - V_T)^2 \left(1 + \frac{V_{DS} - V'_{DS}}{V_A}\right) \quad (D.1)$$

where

$$V_T = V_{T0} + \gamma(\sqrt{\phi_o + V_{SB}} - \sqrt{\phi_o}) \quad (D.2)$$

$$V'_{DS} = \frac{V_{GS} - V_T}{a} \quad (D.3)$$

where a is a process dependent quantity somewhat larger than 1, and all other terms have their usual meanings. For low and medium frequencies, the model for the transistor is shown in Figure D.2. The values of the various quantities in Figure D.2 in terms of the terminal voltages, currents and physical parameters are [16]

$$g_m = \frac{\mu C'_{ox}}{a}\left(\frac{W}{L}\right)(V_{GS} - V_T) \quad (D.4)$$

208 HIGH FREQUENCY CONTINUOUS TIME FILTERS

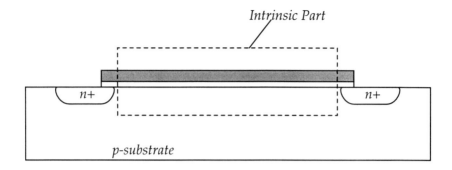

Figure D.1. The intrinsic part of an MOS transistor.

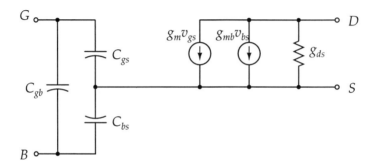

Figure D.2. A simple quasistatic model for the MOS transistor.

$$g_{mb} = \left(\frac{\gamma}{\sqrt{V'_{DS} + V_{SB} + \phi_o} + \sqrt{V_{SB} + \phi_o}} \right) g_m \quad \text{(D.5)}$$

$$g_{ds} \approx \frac{I_{DS}}{V_A} \quad \text{(D.6)}$$

$$C_{gs} = \frac{2}{3} C_{ox} \quad \text{(D.7)}$$

$$C_{bs} = \frac{2}{3}(a - 1) C_{ox} \quad \text{(D.8)}$$

$$C_{gb} = \frac{a - 1}{3a} C_{ox} \quad \text{(D.9)}$$

Appendix D: SMALL SIGNAL MOS TRANSISTOR MODELS   209

Figure D.3.   MOS transistor model in the off condition.

2.   MODEL WHEN THE DEVICE IS OFF

When the device is off, the small signal model reduces to Figure D.3. In that figure

$$C_{gb} = C_{ox} \frac{\gamma}{2\sqrt{V_{GB} - V_{FB} + \frac{\gamma^2}{4}}} \qquad (D.10)$$

3.   EXTRINSIC PARASITICS

The extrinsic parasitic capacitors near the drain of a transistor are shown in Figure D.4. C_{db} is the depletion capacitance between the drain contact and the substrate. C_{gd} is the physical overlap capacitance between the gate and the drain. Similar capacitances exist at the source end of the transistor. An important thing to notice is that C_{gd} and C_{db} stay the same irrespective of whether the transistor is on or off as long as the gate and drain are maintained at the same potential. Figure D.5 shows the complete model for the MOSFET. All extrinsic capacitors have the subscript 'e'. C_{sde} can also account for any wiring overlap capacitance between the source and the drain .

210   HIGH FREQUENCY CONTINUOUS TIME FILTERS

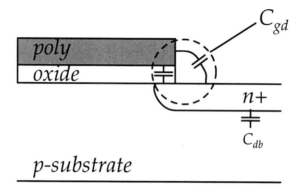

Figure D.4.   Extrinsic parasitic capacitors near the drain.

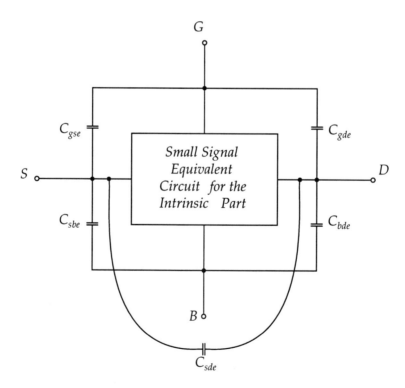

Figure D.5.   Extrinsic transistor capacitances added to an intrinsic model.

Appendix E
CALCULATION OF THE STEADY STATE WAVEFORM OF THE FILTER COMPARATOR OSCILLATOR

For this analysis, the reader is referred to Figure 8.11. We use $T = 2t_1$ from (8.18), and denote by k, the even integer $2m$. Then, from (8.17), we get,

$$y(kt_1 + \tau) = \sum_{n=0}^{k}(-1)^n s(kt_1 + \tau - nt_1), \quad 0 < \tau < t_1 \qquad (E.1)$$

Using a change of variables, and keeping in mind that k is even, for $\tau < t_1$ (E.1) can be written as follows:

$$y(kt_1 + \tau) = \sum_{n=0}^{k}(-1)^n s(\tau + nt_1) \qquad (E.2)$$

In order to avoid unwieldy expressions, and in preparation for the development that follows, we use the notation

$$\omega_{osc} = \omega_0 \sqrt{1 - \frac{1}{4Q^2}} \qquad (E.3)$$

Equation (8.8) for $t = \tau + nt_1$ becomes, using (8.11) and (E.3):

$$s(\tau + nt_1) = \frac{1}{\sqrt{1 - \frac{1}{4Q^2}}} \exp\left(-\frac{n\omega_0 t_1}{2Q}\right) \exp\left(-\frac{\omega_0 \tau}{2Q}\right) \sin(\omega_{osc}\tau + n\pi) \qquad (E.4)$$

Using (E.4) in (E.2), and noting that $(-1)^n \sin(x + n\pi) = \sin(x)$, we get

$$y(\tau + kt_1) = \sum_{n=0}^{k} \frac{1}{\sqrt{1 - \frac{1}{4Q^2}}} \exp\left(-\frac{n\omega_0 t_1}{2Q}\right) \exp\left(-\frac{\omega_0 \tau}{2Q}\right) \sin(\omega_{osc}\tau), \quad 0 < \tau < t_1 \quad (E.5)$$

or, using (8.11):

$$y(kt_1 + \tau) = \frac{\exp(-\frac{\omega_0 \tau}{2Q}) \sin(\omega_{osc}\tau)}{\sqrt{1 - \frac{1}{4Q^2}}} \sum_{n=0}^{k} \exp\left(-\frac{n\pi}{\sqrt{4Q^2 - 1}}\right), \quad 0 < \tau < t_1 \quad (E.6)$$

To calculate the steady state response, we allow k to increase. In the limit, replacing the sum by an infinite sum, and using

$$\sum_{n=0}^{\infty} x^n = \frac{1}{1-x}, \qquad |x| < 1 \quad (E.7)$$

the right hand side of (E.6) becomes

$$\frac{\exp(-\frac{\omega_0 \tau}{2Q}) \sin(\omega_{osc}\tau)}{\left[1 - \exp(-\frac{\pi}{\sqrt{4Q^2-1}})\right] \sqrt{1 - \frac{1}{4Q^2}}}, \quad 0 < \tau < t_1 \quad (E.8)$$

It is obvious from Figure 8.10 that for $t_1 < \tau < 2t_1$, the steady state response shape is the same as in the interval $0 < \tau < t_1$, except for a sign inversion. Thus, (8.19)–(8.22) in Section 4. hold, where T is as in (8.18).

Index

CMOS, 48
Gm-C, 42
Gm-OTA-C, 46, 149
MOSFET-C, 37
Monte Carlo, 102
Nauta's technique, 149
Accumulation, 8, 12
Binary weighted, 98
Biquad, 34
Butterworth filter, 95
   quality factors in a, 95
Capacitors
   Depletion, 118
   Empirical formula for interconnect, 118
   Interconnect, 118
Constant-capacitance scaling
   Implementation of, 91
Constant-conductance scaling
   Implementation of, 90
Crossmodulation, 87
Degeneration, 49
Depletion, 28
Disc-drive, 1
Distortion in scaled networks, 87
Distortion, 7, 31, 86
   Measured filter, 143
Dynamic range, 56, 89
Excess noise factor, 50, 53–54
Excess phase, 65
Feedthrough, 133
Flatband voltage, 18
Gate overdrive, 103
Integrator, 34
Intermodulation, 87
Inversion, 8

Layout
   Balanced, 120
   Biquad, 120
   Offsets due to, 101
Mismatch, 100
   current factor, 100
   oxide-thickness dependence of, 100
   threshold voltage, 100
Mixed nodal analysis (MNA), 80
Noise in constant-conductance scaled
   networks, 76
Noise spectrum, 138
Nonlinear capacitor, 6
Passband, 136
Phase error, 34
Polysilicon gate depletion, 19
Programmability, 56
Quality factor, 28, 33
Read-channel, 1
Right half plane zero, 38
Scaled networks
   Noise properties of, 75
   Nonlinear, 79
   Properties of, 73
Scaling, 72
   constant-capacitance, 74
   constant-conductance, 74
Scattering parameters, 135
Surface potential, 10
Transconductor, 98
   Modified Nauta, 152
   constant capacitance scaled, 92
   gain of, 98
Transconductors, 48
Triode operated MOSFETs, 51